Implementing the NCTM Standards

A Bridge to the Classroom

Implementing the NCTM Standards

A Bridge to the Classroom

Grades 5-8 and 9-12

Margaret A. Farrell

Janson Publications, Inc.

DEDHAM MASSACHUSETTS

Acknowledgement: Chapter openers for Chapters 1 through 5 are reproduced from *Curriculum and Evaluation Standards for School Mathematics*, © 1989, National Council of Teachers of Mathematics, and appear here with permission of the publisher.

Printed in the United States of America.

987654 123456789

Implementing the NCTM *Standards*

PRELIMINARIES

CONTENTS

PREFACE

This text is designed to form a bridge between the vision presented by the *Curriculum and Evaluation Standards for School Mathematics* (National Council of Teachers of Mathematics, 1989, hereafter referred to as the *Standards*) and the classroom. The structure of the bridge is based on the instructional and evaluation principles found in *Secondary Mathematics Instruction: An Integrated Approach* (Farrell and Farmer, 1988). This text is directed towards middle and high school mathematics teachers—the population of teachers designated as *secondary*. The authors of the *Standards* divide the grade levels into three grade ranges: K–4, 5–8 and 9–12. In this text, it was decided to use the 5–8 and 9–12 groupings to represent the middle and high school years. These are, in fact, the same groupings that are addressed by the term *secondary* in the title of the Farrell and Farmer text.

This text might be used in a methods course, in seminars associated with a student teaching course, or in a subsequent mathematics education course at the undergraduate or graduate level. It would also be appropriate to use it in an inservice course directed toward better understanding of, and implementation of, the *Standards*.

The guidelines in the *Standards* have been, and continue to be, the baseline indicators illustrating the kind of mathematics learning needed in our schools. The authors of the *Standards* and the leadership of the National Council of Teachers of Mathematics have encouraged classroom teachers at all levels to illustrate ways to apply the message of the *Standards* in articles in the NCTM journals and in sessions at national, regional, and local meetings. A commission of the National Council of Teachers of Mathematics and its working groups developed a complementary document, *Professional Standards for Teaching Mathematics* (NCTM, 1991), to further clarify the vision of the *Standards*. In the *Professional Standards* monograph, the authors focused on standards for the kind of teacher knowledge, beliefs, and classroom strategies needed to promote the message of the original *Standards* document, as well as standards that speak to the kind of teacher evaluation and support mechanisms needed to encourage reform. The National Council of Teachers of Mathematics also commissioned a set of addenda to the *Standards* at each of the three grade ranges. These resource books include activities and investigations that might be used by teachers to address one or more of the standards at a particular grade range.

In some school districts, inservice workshops have been designed to help mathematics teachers implement these new teaching ideas. The challenge is even greater for preservice teachers, whose classroom roles have been primarily those of learners. Thus, it is imperative that instructors of methods courses help their students understand the vision embraced by NCTM in the *Standards*, and at the same time assist them in moving from student roles to roles as teachers of mathematics. The latter is no simple transformation.

Methods students and preservice teachers are just beginning to learn about the connections between objectives and strategies. Now, in addition to re-examining their own mathematical knowledge and analyzing the connections of one formula or definition to another, they must also examine the newest approach to curriculum and evaluation, the *Standards,* and consider the following questions.

> What are the curricular priorities emphasized in this document?
>
> How are they different from those they have been exposed to, as students?
>
> If they keep those priorities uppermost in their planning, what are the implications for strategy design, for testing, for diagnosis and feedback, for choice of materials?
>
> How can the principles they have learned from a methods class be applied to achieve these curricular priorities?

At the beginning of this Preface, the text was characterized as a bridge between the vision of the *Standards* and the instructional and evaluation principles needed to implement that vision. An examination of the Contents shows how the bridge concept was incorporated into the text. For each standard considered here, there is an analysis of the standard itself with attention to possible misconceptions and areas needing special emphasis. Next is the *Bridge* section with relevant explanatory or illustrative material.

Throughout the text, the reader is directed to specific pages in *Secondary Mathematics Instruction* for further study. The instructor is the best judge of whether the students need to examine other references for topics such as research on learning styles in mathematics, conditions for learning concepts or rules, *etc.* If there is a need and the students are using a methods text other than *Secondary Mathematics Instruction* , the instructor can direct their attention to appropriate sections in their methods text. Instructors who prefer to use their own notes in methods classes may wish to have a few copies of *Secondary Mathematics Instruction* available for student reference, as needed.

Sets of activities are included in this text as support material for the methods instructor, student teaching supervisor, or inservice workshop leader. These may be used to help college students construct strategies, analyze materials, and explore new concepts. Many of the activities have been designed as open-ended questions, to which there might be several different and reasonable responses. The kind of interaction that occurs when students discuss different responses and provide justifications for these responses is consistent with the classroom environment promoted by the authors of the *Standards*. It is also consistent with the kind of preservice and inservice environment spoken of by the authors of *Professional Standards for Teaching Mathematics.* Instructors are the best judges of when to assign these activities and whether to use them for in-class small group work or for individual or shared out-of-class assignments.

This book consists of two sections. One of these contains the text itself. The other, called the Reader, contains specific lesson ideas, plan outlines, and evaluation techniques excerpted from *Secondary Mathematics Instruction.* Many of the Activities, as well as questions posed within the text, are based on one or more of these Readings. In this way, all users of the text have access to the same set of idea materials. If you have been using *Secondary Mathematics Instruction* as a methods text, you will find that the questions posed in this text go beyond those posed in the earlier text and afford the students another opportunity to reflect on the problem-solving nature of teaching.

Another perusal of the Contents shows that exactly four of the curricular standards are studied in this text. These four are *Mathematics as Problem Solving, Mathematics as Communications, Mathematics as Reasoning,* and *Mathematical Connection.* Why did the author choose to analyze only four of the curriculum standards and why these four? An examination of the thirteen standards at each of the three grade groupings shows that these four standards are listed as the first four at each level. They are identical in form and general intent, although illustrated in increasingly more sophisticated ways as one moves up the grade levels. These four standards are the glue that unifies all thirteen standards in a particular grade grouping and promotes the articulation so necessary across grade groupings. Each of the other thirteen standards is couched in specific content terminology, but the first four are generic and exemplify the dynamic view of mathematics that permeates the pages of the document. Thus, these first four standards are considered in some detail since a curriculum without their presence would bear no resemblance to that framed by the authors of the *Standards*.

In order to use this text most profitably, it is important for the students to have reviewed some, or all, of the *Standards*. It is also desirable that the methods student, student teacher, or inservice teacher already have some background in the readings and activities in *Secondary Mathematics Instruction*, or some comparable text. In the long run, the instructor is the best judge of when and how to interweave the materials for a particular course and a specific group of students.

Given the importance of teacher beliefs and knowledge in the design and implementation of classroom lessons, it is essential that teachers are helped to reflect on all aspects of teaching mathematics. This text's most important function might be as a catalyst that fosters such reflection and thus helps preservice and inservice teachers integrate research, practice, and theory.

TO THE STUDENT

This text is designed to help you make connections between previously learned theory and practice, classroom experiences you have had or will have as a teacher or student teacher, and the vision embodied in *Curriculum and Evaluation Standards for School Mathematics* (the *Standards*).

The *Standards* presents a vision of the kind of mathematical understanding to be demonstrated by all students. In a *Standards* classroom, an observer would note an active learning environment in which students are constructing mathematical meaning under the guidance of the teacher, rather than one in which students are passively receiving information from the teacher. The authors of the *Standards* concede that achievement of the goals they have set is not an easy task. Inservice teachers may need to learn some new strategies, modify their approach to certain curricular objectives, and re-examine the nature of some mathematical concepts. If you are a beginning teacher, the task is different. You are already starting to rethink mathematical understandings from the point of view of the learners you might face. Simultaneously, you are learning how to transform yourself from student to teacher—a teacher who is a facilitator rather than a presenter. This text is designed as a bridge between the *Standards* and the classroom—to help you identify instructional and evaluation principles that can be applied to the design of strategies, the testing of objectives, and the selection of resource material.

"So I'm supposed to worry about playing a part at the beginning of some new reform movement," I can almost hear you saying. "I'm more worried about next week's test and how and if I'll survive in the classroom when I start student teaching." I completely agree with you. Survival, both on the way to and in the mathematics classroom, is your first concern. Believe me, it's an important reason for the existence of this text.

One of the goals of the text is to give you a new opportunity to reflect on important ideas and principles—not necessarily so that you can ace a test, but so that you can survive in the mathematics classroom. However, the concept of survival I have in mind is more powerful than that of just getting you through the day. Sometimes we visualize survival as desperation of the kind that must be felt by a ship-wrecked person alone in a life raft. The concept of survival I have in mind is expressed best by the ebullient statement of someone who has just been through an arduous or dangerous experience: "I'm a survivor!" That hosanna-like declaration carries with it a sense of joy and life. In the case of mathematics teaching, it's the joy and life you infuse into the mathematics you teach.

Why should reflecting on what you've already studied be of practical help in the classroom? Perhaps if teaching were a motor skill, like routine and repetitive machine work, there would be less payoff from studying and reflecting on your actions. However, you already know that teaching is a complex activity, a problem-solving activity of the highest order. In a recent case study of three mathematics teachers, Feldt (1991) observed and interviewed these teachers over a five-week period and found that they consistently reflected upon, and evaluated, their own day-to-day strategies in an attempt to make mathematics meaningful for their students. While one study does not prove the point, these teachers demonstrated that, for them, effective teaching required ongoing self-evaluation or reflection. They consistently reflected on answers to questions such as

> What happened in this lesson?
> What could I have done differently?
> What must I do tomorrow?

The answers to these questions require observation, reflection, and analysis.

In the text, one chapter is devoted to each of the first four standards listed in each of the three grade ranges. These four standards, *Mathematics as Problem Solving, Mathematics as Communication, Mathematics as Reasoning,* and *Mathematical Connections*, were selected because they are the conceptual glue that holds together all of the other standards. For example, consider the first standard, *Mathematics as Problem Solving.* The student must apply problem-solving techniques and have a questioning attitude in the specific content areas referred to in the other nine standards. Similarly, Standards 2, 3, and 4 also infuse the other nine specific content standards. Another focus in this text is on the students enrolled in the grades often characterized as middle and high school. The authors of the *Standards* subdivided their work into three groups of grades. The upper two, 5–8 and 9–12, are the ones considered in the text since these are the grades included in most designations of secondary school. In the fifth chapter you are asked to consider related evaluation standards from the *Standards,* while the final chapter is devoted to issues highlighted by the current reform movement.

You will find that this text is written in an interactive fashion to encourage your reflection. However, whether that reflection occurs depends entirely on you. Try to respond to questions before reading further. When you are asked to share data and conjectures with a colleague, look on the request as a way to gain additional insight into the topic being considered.

The principles in this text are based on those delineated in *Secondary Mathematics Instruction: An Integrated Approach.* Thus, there will be references to pages in that methods text where background information on a topic may be found. If you are not using that text, or do not have access to a copy, your instructor may direct you to appropriate background material in another source.

Notice that this volume has two sections. The second section, the Reader, has been perforated so that you may separate it from the text and thus have the two sections open at related pages simultaneously. The Reader contains excerpted lesson plan ideas, transcripts of class activities, or other specific ideas from *Secondary Mathematics Instruction*. The other section contains the main body of the text. Throughout the text, you may be asked to read a section from the Reader and then respond to questions based on the material in the excerpted section. If you have already used *Secondary Mathematics Instruction* as a text, you will find that the questions in this text take you beyond those in the earlier one.

The activities in this text are designed to help you make sense of the meaning of some standard or ascertain the advantages or disadvantages of using a particular strategy to reach a standard. The proof of the pudding, in this case, is the success you will have as you effectively implement the principles in the text to attain the vision of the *Standards*.

It is hoped that this text will serve as a catalyst to further help you reflect on the nature of mathematics teaching and learning. This kind of reflection may be the most important prerequisite to survival as an effective mathematics teacher.

M. A. Farrell

Implementing the NCTM *Standards*

THE TEXT

Curriculum Standards for School Mathematics

STANDARD 1: Mathematics as Problem Solving

In grades 5–8, the mathematics curriculum should include numerous and varied experiences with problem solving as a method of inquiry and application so that students can

- *use problem-solving approaches to investigate and understand mathematics content*
- *formulate problems from situations within and outside mathematics*
- *develop and apply a variety of strategies to solve problems, with emphasis on multistep and nonroutine problems*
- *verify and interpret results with respect to the original problem situation*
- *generalize solutions and strategies to new problem situations*
- *acquire confidence in using mathematics meaningfully*

In grades 9–12, the mathematics curriculum should include the refinement and extension of methods of mathematical problem-solving so that all students can

- *use, with increasing confidence, problem-solving approaches to investigate and understand mathematical content*
- *apply integrated mathematical problem-solving strategies to solve problems from within and outside mathematics*
- *recognize and formulate problems from situations within and outside mathematics*
- *apply the process of mathematical modeling to real-world problem situations*

Reprinted with permission of the publisher. See acknowledgements.

Chapter One

Mathematics as Problem Solving

Problem solving is such a familiar goal of mathematics teaching that it's tempting to dismiss it as trite. You can find explicit references to the importance of reflective thinking in the solution of problems in a 1938 report, *Mathematics in General Education.* In the half-century since that time, there have been numerous articles, lectures, and books by mathematics educators on ways to teach problem solving. Surely, given all this attention to the teaching of mathematical problem solving, teachers know how to teach for it and test for it. Don't they? The answer depends on the teacher's definition of mathematical problem solving and of what constitutes a problem. The authors of the *Standards* addressed this issue in their explanation of the first standard, **Mathematics as Problem Solving**. They found problem solving sufficiently important to list as the first standard at all three grade groupings. Thus, **Mathematics as Problem Solving** was identified as more than a goal. Rather, it is really a unifying theme for the curriculum throughout each group of grades and across grade groupings. As you read the descriptive material under other standards, you will find that the problem-solving standard is related to the other standards and serves as one of the four unifying themes of the document.

How is a standard, or goal, defined or described by the authors of the *Standards*? Two techniques are used. First the standard is broken down into a list of sub-goals, or outcomes. Then descriptive material containing illustrative classroom scenarios is provided. If you haven't already read the Standard 1 sections for grades 5–8 and 9–12, you may wish to stop here and check on the material provided in those sections. The list of sub-goals or outcomes is reproduced on the opposite page for reference.

The Vision: *What Is a Problem?*

Strange as it may seem to dedicated problem-solvers, the first stumbling block in the design of a problem-solving lesson is the common confusion surrounding the term *problem*. Secondary students have typically been exposed to the use of the term as a routine exercise, a type word problem, or a standard proof.

What meaning is being applied by the writers of the *Standards* when they use the term *problem solving*? Reread the list of student outcomes cited under Standard 1. I'm sure you'll agree that, on both the 5–8 and 9–12 grade levels, the nature of the problems alluded to is neither routine nor typical in nature. For instance, the problem must be one that provides an opportunity for conjecture and exploration, or for synthesis of several known approaches. In the grades 9–12 set, the authors have built in a student outcome related to the application of mathematical modeling to real-world situations. That kind of learning hardly seems possible if all of the questions are routine or typical.

In order to achieve the Standard 1 outcomes, the problems under consideration cannot be practice examples or even type problems of the kind so necessary to the sciences (*e.g.*, lever problems). That doesn't mean the authors of the *Standards* intended that practice exercises should be eliminated or that type problems should be deleted. Without the ability to recognize important types and apply learned algorithms to their solution, the student would lose potentially useful approaches to non-standard, or novel, problems.

Where are outcomes related to important formulas, necessary type problems, *etc.* found in the *Standards*? Review the other nine standards for grades 5–8 and 9–12. Which of these are likely to be related to the need to learn algorithms, formulas, and other forms of rules? Consider, for example, Standard 6: **Number Systems and Number Theory** in the grades 5–8 section. Identify an algorithm indicated by an outcome such as

> extend their understanding of whole number operations to fractions, decimals, integers and rational numbers.

You may have listed the addition algorithm for any of the classes of numbers mentioned in the outcome; or the subtraction, multiplication, or division algorithms. However, the authors chose their words carefully. The verb *extend* means to develop these algorithms as extensions of one another, rather than as distinct rules. Similarly, an examination of the outcomes under Standard 8: **Geometry from an Algebraic Perspective** from the grades 9–12 section leads to the identification of multiple and essential algorithms to be learned. In order to further familiarize yourself with this aspect of the *Standards*, try your hand at the activities that follow.

Activities

Refer to the *Standards*.

1-1. Choose one standard from Standards 5–13 at both the 5–8 and 9–12 grades. Identify some formulas or algorithms that either directly, or by implication, would need to be learned to reach one of the outcomes under each of the selected standards.

1-2. Review the standards you selected in Activity 1-1. Are there some algorithms or type problems that would either be de-emphasized or deleted from consideration? If yes, what are they and what rationale is given in the *Standards* for this action?

A second, more subtle, issue related to choice of problem is associated with the difficulty level of a question. Since, in daily life, we so often use the term *problem* to characterize a difficult decision or an arduous task, it is tempting to select as problem contexts those exercises that have been found difficult by many students—perhaps algorithms that seem to attract systematic errors. Again, there is a significant learning issue represented by such problem types. However, in these cases, the desired outcomes for the students are error-free performance and ability to reduce a complicated typical problem to a simplified form. The solution of such type problems does not test outcomes such as the ability to make reasonable conjectures, to restructure a situation by temporarily ignoring one of its characteristics, or to construct a question from a collection of data.

The importance of these process-oriented outcomes cannot be underestimated. Long after specific formulas and definitions have been forgotten, confidence in one's ability to interpret and work toward the solution of an unfamiliar problem (*e.g.,* the repair of a car or a bicycle, the assembly of a toy or piece of furniture, recreational boating problems, or gardening decisions) is the long-term transfer of mathematical learning promoted by Standard 1. Moreover, for adults in a wide range of careers (*e.g.,* nurse, research scientist, construction worker, engineer, set designer, or architect), these process-oriented outcomes are essential.

A Bridge to the Classroom

There are important differences in the way strategies for solving type problems versus those for novel problems are learned, and thus in the way each is taught. This doesn't mean that type-problem lessons need to be static and low in interaction, while novel-problem lessons are vibrant. What are these differences and how does a teacher promote the kind of atmosphere alluded to in the *Standards*, whether teaching a lesson dealing with type problems or with novel problems?

In order to answer these questions, we will examine the defining characteristics of type problems and the instructional strategies associated with the teaching of type problems. Then, we'll reconsider novel problems and the instructional strategies associated with the teaching of novel problems.

Type Problems

To begin with, let's study an outcome statement always present in type-problem lessons.

> ... the outcome is for the students to identify a problem as belonging to a certain type, a type of sufficient interest and usefulness that it recurs frequently in the real world. (Farrell & Farmer, 1988, pp. 175–176)

Perhaps the bad name associated with lessons on type problems is due to the **kind** of type problems found in those lessons. For instance, it is difficult to characterize most age problems as questions that meet the criteria of usefulness. Furthermore, even when the selected type problems are of eventual practical value, sometimes the context is uninteresting and meaningless to the students. There has been more than one comedy routine based on supposed real-world problems that never saw the light of day in the real world. Do you remember motion problems about two trains rushing towards one another at differing speeds? The question in the text might have been: *How long before they reach the same location?* The comedian would have substituted the word *collide* and wonder aloud about the future of engineers who set those trains on a collision course. In contrast, a question might be asked about the results of a cross-country race between two school teams.

Table 1.1 Results from a Cross-Country Race between Rangers and Scouts

Finish	Team	Time in Minutes
1	Rangers	15:20
2	Scouts	15:27
3	Scouts	15:29
4	Rangers	15:57
5	Rangers	15:58
6	Rangers	16:00
7	Scouts	16:01
8	Scouts	16:02
9	Rangers	16:10
10	Scouts	16:11

The score for the Rangers is 25, calculated by adding the finishes of the team members $(1 + 4 + 5 + 6 + 9)$. The score for the Scouts is 30. Since the winning team is the one with the lowest rank sum, the Rangers won this event. However, suppose that the winner was declared on the basis of average time per team. Would the Rangers still win?

The cross-country race question has intrinsic interest because it is an event familiar to secondary school students. The possibility of using a different method to identify the winning team would also be inherently interesting, and could quickly lead to subsequent discussions on the way in which winners are chosen in other events (*e.g.,* Olympic trials).

The next important characteristic of type problems is the algorithmic nature of their solution. Certain word problems in the sciences are known by the formula associated with their solution (*e.g.,* Ohm's law), while some have been reduced to a kind of standard procedure (*e.g.,* solution or mixture problems). As mathematics teachers often like to tell students, our subject helps us to find shortcuts. In the case of solving type problems, the very nature of the task indicates that an outcome is to identify not only the type, but the corresponding model solution procedure, and to learn how to apply that procedure in diverse cases.

That is why it is inappropriate, and confusing, to begin the teaching of a type problem by allowing the students to believe that diverse conjectures and diverse solution methods are desired **final** outcomes. When the teacher completes the lesson by allowing only one kind of solution process, stated in one agreed-on way, those students who tried to devise creative approaches become disillusioned and are more likely to join the chorus of "Just tell us how to do it so we can copy it down" the next time the teacher tries to encourage them to construct original approaches.

Before we examine how type-problem lessons can be designed to encourage interaction, despite what may seem to be a ban on creativity, first check on your understanding of the nature of type problems. Remember, a question becomes a type problem if the desired outcomes are:

> to analyze and identify its characteristics in order to use these characteristics to recognize other questions in this same family

and

> to learn a basic solution strategy that may be adapted to all other questions in this same family.

Activities

1-3. Return to question 1-2 from the first set of Activities and the standards selected by you in your response. Would you delete any other type problems or algorithms on the basis of the criteria found in the reading in this section? Why or why not?

1-4. Choose two different problem types that seem to be included in the curricular outline of the *Standards*. For each, write at least one question that satisfies the criterion of being of interest and meaningful to students at that grade range.

Hint: Think of unfamiliar or puzzling things, events, or ideas; the use of color, sound, or touch; situations that students deal with outside of mathematics class, such as after-school jobs, hobbies, or science projects. For more ideas, see pages 189 and 191–192 in *Secondary Mathematics Instruction*.

Instructional strategies

Just as the mention of type problems seems to result in an immediate negative reaction from many adults, lessons that focus on the teaching of type problems are often dull experiences with the final outcome being the memorization of a chart or a procedure. However, if the problem type is interesting and potentially meaningful to the student, and if strategies are designed to match the conditions for rule learning and to correspond to student cognitive development and learning styles, a stimulating lesson depends solely on the teacher's implementation.

On pages 127-128 of the Reader, Mrs. Armstrong outlines a lesson designed to teach high school students how to recognize and solve a particular kind of solution problem. Read the instructional strategies she uses and identify those directed at helping students recognize the pattern of the solution. Pay special attention to the teacher demonstration. Notice how this visual and changing display is to be used as a stimulant for focusing student observations.

Let's return to the part of the lesson in which Mrs. Armstrong helps the students to focus on the desired pattern. In which aspects of the planned lesson does this happen? Check out your ideas with another classmate working on this section. Did you identify the carefully structured demonstration, the importance of the kind of teacher questions, and the selection of a model problem that corresponds closely to the demonstration? Mrs. Armstrong does not intend to begin by stating a rule or drawing a box to be filled in, followed by an explanation. Thus, even though the final goal is for all to reach the same generalization, the process involves students reasoning from given situations and stating ideas in their own words. Notice that the strategy pattern also includes the completion of a model problem by the teacher, with the help of the class, and supervised practice of the generalization.

If the situation is a type important enough to learn, then the actual rule (the algorithmic process, application of the formula, or solution of the word problem) must be practiced in a meaningful way in order to be remembered. Thus, delegating the bulk of the practice examples to homework is not a sound idea. Why not? What are the pitfalls? In the case of type problems, this practice need not always include the complete solution of the problem. In order to have sufficient practice in the actual type being learned, the teacher might include a selection of problems of the same, general type but differing in some key aspect. If students are required to solve completely only one or two of these specific problems, and told to draw the picture and write the equation that would be used to solve the others, they have an opportunity to see if they can apply the learned pattern in different situations.

> Hint: When diagrams modeling the situation are used in the development of a lesson, students should be encouraged to draw similar diagrams when they are working on new problems. Encouraged; not required ! Some students retain these images in their memory or transform them in ways helpful to them. Many others—perhaps surprisingly, some of the highest achievers—find that the drawing of the situation reduces the problem to a simple one, to which they can easily apply the basic solution strategy for the problem type.

Activities

1-5. Ask an inservice teacher to list the three or four formulas or algorithms that their students seem to have the most trouble remembering correctly. Share the list with classmates and analyze the common errors made by students.

1-6. Read the parabola activity on pages 129-130 in the Reader. If the purpose of this activity were to lead geometry students to the locus definition of a parabola, an example of a rule, what components of the activity might promote retention of this rule?

1-7. Review this section for the characteristics of a potentially stimulating type-problem lesson. In what specific ways does Mrs. Armstrong (See Reader, pages 127-128) try to incorporate these characteristics into the solution problem lesson?

Novel Problems

Have you heard the argument that the first time the students encounter a motion problem, it is a novel situation for them? The next step in the argument is that distinctions between typical and novel are really in the eye of the beholder. This argument, though, returns us to the issue of the eventual goal of the teacher for this kind of problem and the need not to mislead the students. In the case of motion problems, the goal is probably **not** to have students construct multiple solution procedures, but to lead them to the eventual application of some form of the distance/time/rate relationship. However, suppose the students are given a problem such as the design of a container to hold a collection of odd tools, with the proviso that some consider the importance of packing many of such containers efficiently; others consider the shape that might be most expediently opened and closed again, as a temporary storage box; and still others identify the problems of making a particular shape given cardboard versus a malleable metal, or plastic versus glass. In this case, the goal is definitely **not** for all to reach some common algorithmic solution that may be repeatedly applied to other like problems. Instead, this kind of problem addresses the goals listed in Standard 1.

Thus, more than any other factor, a problem is useful for a lesson on solving novel problems if it does not belong to a class of problems and if the ultimate goal is to engage in various search and/or restructuring strategies, rather than to find a common solution pattern. The problem may or may not be solved. If it is a real-world problem, it may not be possible to do more than agree on a range of possible approaches. On the other hand, there may be in fact a single numeric or algebraic answer to the problem, but the nature of the problem must be such as to allow for a variety of approaches and a period of search behavior. Some of these problems lend themselves to long term projects—projects in which all approaches, including false starts with rationale, are recorded in log form. Students could be advised that these logs will be discussed in small groups by a specified date, even if a final solution has not been reached. (See pp. 113 and 138–140 in *Secondary Mathematics Instruction* for more background on the learning of novel-problem solving.)

The Activities that follow provide generic ways to manufacture non-typical problems as material for small group discussion in almost any topic area.

Activities

1-8. Prepare a handout with the **answers, but not the questions** in a topic area of a mathematics course from grades 5–12. Use the following as your directions:

*What you see here are answers to questions from the unit on
_____. For each answer, devise one or two appropriate
questions. Be sure to check your response by working out the
solution to your questions. Your results will be shared with
those of others in small group work tomorrow.*

1-9. Prepare a handout with typical questions from a unit in a
mathematics course from grades 5–12. In addition, provide an
erring student's (Careless Charlie's or Careless Charlene's)
"solutions." Use the following set of directions:

*The questions and solutions that follow were copied from CC's
paper. In each case, CC has made one or more errors, even
when the correct final answer was found! Your job is to find all
errors, correct the entire solution, and be prepared to explain
CC's misconceptions that led to these mistakes. You will be
asked to share your responses with those of others in small
group work tomorrow.*

(This handout could start out as a homework assignment with small
group discussion in class. Secondary students could also be asked to
devise an explanation that might help CC understand the mistaken
procedures.)

1-10. Choose a topic area and ask students to generate their own
problems. They will need some help in doing this since their first
efforts are likely to be copies of typical problems with changes in
numbers. A good way of assisting students is to provide them
with some data. These can be graphs from the newspaper,
sports statistics, a statement on politics or famine, the TV ratings,
data on drug use, *etc.*

Instructional strategies

What effect do the goals associated with novel-problem solving have
on instructional strategies? If the students are to work on the
problems in class, they would profit from working with one another—in
pairs or in small groups of four or even five. What better way to help
them to investigate, conjecture, and try a variety of tactics? You only
need to imagine a classroom in which the teacher has the central role,
guiding the students with questions and cues, and it becomes
obvious that only the rare student would actually be constructing.
Most, under the best of circumstances, would be following along and
occasionally volunteering. Thus, for the goals in Standard 1 to be
realized, there must be a healthy dose of students talking to, and
working with, students. These discussions may take place in the
context of cooperative learning groups already being used by the
teacher.

In Davidson (1990), a high school mathematics teacher reported
how he transformed almost all his instruction into cooperative learning
groups. Alternatively, small groups may be formed to fit specific
lessons, such as the one designed by Mr. Potter for his geometry

class (See page 131 of the Reader). Study the instructions he gives his students and then return to this page.

The kinds of problems being posed are proofs and, from the instructions, it would appear that Mr. Potter expects the students to be able to solve these without too much trouble. What feature saves this material from simply being a practice session on typical proofs? If you keyed in on the teacher's expectations for the lesson, the importance of diverse strategies, and the subsequent analysis and evaluation expected from the students, you have correctly identified the way Mr. Potter builds novelty into his lesson. His preliminary comments before the instructions identify the kind of groundwork needed before a teacher can hope to enjoy success with this lesson. His directions to the class show the kind of organizational planning that is needed if the small groups are to work efficiently.

Activities

Answer the following questions on the basis of Mr. Potter's lesson material.

1-11. When are the students actually asked to move into groups? React to the timing of the move.

1-12. What kinds of information does Mr. Potter give the students while they are still in a whole class organizational arrangement? Why?

1-13. How does Mr. Potter expedite the small group organization? For example, how does a student find out what group he/she is in?

1-14. For a hypothetical class of 19 students, design an overhead master that might correspond to that referred to in Mr. Potter's instructions.

1-15. What other kinds of decisions did Mr. Potter have to make in preparing his overhead master?

Mr. Potter's organizational planning might appear overly structured given that he wants to establish an atmosphere in which students can feel free to bounce ideas off one another. However, it is a truism that the more open and free-wheeling the classroom environment becomes, the more possibilities exist for ineffective use of time and talent. Thus, in order to make such an environment reach its potential, the teacher must be well-organized, communicate the purpose of the work clearly, and be a vital part of the ongoing classroom work.

Hint: If you are trying small group work for the first time and the students have never experienced small group work in mathematics class, you will need to share your expectations with them. If you are student teaching, ask your cooperating teacher to help you structure small groups to avoid problems. For instance, where should you seat two talkative chums? It helps to prepare a discussion handout in sufficient quantity for all students. Some teachers like to include a response sheet so that all students must write individual ideas, as well as group results. (Some additional suggestions may be found on pages 11–15 in *Secondary Mathematics Instruction.*)

As you might expect, small groups can be working on manipulatives and talking about patterns they have found. That happens in the Primes lesson plan, reproduced on pages 132-136 of the Reader. Read the plan and pay particular attention to the teacher's stated outcomes and the way the checker lab is used. It helps to "work through" part of the lab with pennies (if checkers aren't handy) so that you can visualize the results students would get.

Notice that the Primes lesson is not a problem-solving lesson, but a lesson designed to present concepts. Thus, small groups can certainly be used in lessons other than novel-problem solving! In both cases, Mr. Potter's lesson and the Primes lesson, the teachers eventually get the attention of the entire class, get ideas from groups, and summarize. However, there are two distinct differences in the two lessons. The first has been noted before. Mr. Potter makes it clear that the outcomes are to find diverse solution processes and to analyze and evaluate these processes, rather than to find a particular solution to a question. What does the teacher tell the student pairs in the Primes plan? What are they being asked to find, as a final result?

The second difference in the two lessons is in the nature of the summary that occurs. Study this excerpt from the Instructional Strategy Pattern for [Novel] Problem Solving (*Secondary Mathematics Instruction*, page 183): "...*reassemble class; ask students to weigh the advantages and disadvantages of the proposals (including processes) that resulted from the group discussions*...." Notice the nature of the desired summary. In contrast, in the Primes lesson, the summary takes the form of finding out if the students can generalize the concept to cases not previously tested. Even more dramatic is the difference between the summary in a lesson on solving novel problems and the summary in a lesson on solving type problems. In the latter, the application of the final rule or formula or agreed-on algorithm is the focus. For novel problems, the process is the focus. The next Activity provides questions designed to further clarify these differences.

Activity

1-16. Read about kite lessons on page 137 of the Reader.

a. Try your hand at deducing properties of a kite, given the information in the selection. Construct one or two definitions for special kites and use these new definitions to deduce further theorems and corollaries.

b. On the basis of your response to **a**, answer the following questions:

In what ways could these lessons lead to the kind of student outcomes delineated in Standard 1? What questions would the teacher want the entire class to consider with respect to student construction of definitions and postulates?

So far the strategies outlined here have been designed around in-class activities, but problem-solving strategies might well be based on out-of-class activities or research. Indeed, since one of the characteristics of the successful problem solver is persistence, it is important that time constraints not always be limited to the length of a class period. Long-term homework assignments and individual project work that are designed to have students reach the outcomes listed under Standard 1 are two very useful instructional modes. In these cases, students **may** talk to other students or interview some adult, or they may work in the library, try out ideas on a computer, or design and build a concrete model to verify a conjecture. Thus, the kinds of mental activity associated with novel-problem solving can occur outside of class as well as individually.

> Hint: There are two keys to the successful use of such modes to achieve problem-solving outcomes. The question, or problem, posed by the teacher must be designed to continue to motivate students working over a long term, or working without immediate feedback from classmates and the teacher. Then, when the assignment time has ended, it is important for there not only to be written feedback from the teacher, but also some sort of in-class sharing of results. Again, the emphasis for all students is on the processes used.

Activities

1-17. One atypical homework exercise is to have the students interview a sample of people to elicit their beliefs on a mathematics-related topic.

a. Use this suggestion to design a specific homework assignment for students in a particular mathematics course.

b. Identify the student outcomes related to Standard 1 that are likely to result from completing this assignment.

c. Outline a way to share student results with the class and identify the probable processes you would want to highlight.

1-18. Review the sample projects on pages 138-139 of the Reader. Choose one project. What student outcomes related to Standard 1 would it be reasonable to expect as a result of student completion of the project?

In the reading and the activities of this chapter thus far, the separate outcomes under Standard 1 have not been isolated for consideration but consolidated under the rubric of **Mathematics as Problem Solving.** However, one of those outcomes—that dealing with mathematical modeling—deserves special attention. The statement about mathematical modeling appears as the last outcome under the grades 9–12 group. Is there an analogous statement in the grades 5–8 group of outcomes? What might the authors of the *Standards* have had in mind when they included this outcome?

The Vision: *Mathematical Modeling*

Not only is mathematical modeling noted in one of the student outcomes from Standard 1 (grades 9–12), but it appears explicitly in a student outcome under Standard 6: Functions (grades 9–12) and Standard 7: Geometry (grades 9–12) and implicitly in the outcomes listed under many of the other standards. Moreover, mathematical modeling is listed as a student outcome at the 5–8 grade level (see the sections on Standards 2 and 4 later in this text and Standards 11 and 12 in the original source).

What did the authors of the *Standards* intend when they used the phrase *process of mathematical modeling*? The answer is to be found in a schema of the process of mathematical modeling, depicted on page 138 of the *Standards*. The schema diagrams the connections from real-world situations to well-formulated questions, to the translation of these questions into mathematical models, and eventually back to the real world and an answer to the original question.

What did the NCTM authors mean by the term *mathematical model?* One answer is found by considering some of the examples they give. An equation is one example. A geometric model that can be used to provide a solution to a famous probability problem (NCTM, 1989, pp. 138–139) is another. Both of these examples are instances of representing real-world data or phenomena in some promising mathematical form. Why, promising? It is important to choose a geometric representation or equation format that can be transformed mathematically so that a solution to the real-world questions can be found.

Why is the process of mathematical modeling cited as a student outcome under the standard associated with problem solving? What understandings are necessary in order that students may construct mathematical models? What are some of the important instructional implications of teaching directed towards this outcome? These are some of the questions that are addressed in the next sections of this chapter.

A Bridge to the Classroom

Take another look at the schema picturing the application of mathematical models from the *Standards*. Then consider the Model of Mathematics (Figure 1.1) on page 15.

The curved arrow in the Model of Mathematics shows the problem solver moving to the idea world of mathematics where the reduced data are idealized (*e.g.,* transformed symbolically into an algebraic expression or equation, or transformed geometrically into a representative shape). Another way of describing this activity is to say that the problem solver is simplifying the real-world situation and transforming it into a mathematical model. The resulting model may be a formula or some other equation, a geometric shape of some sort, or even a graph.

The next step in that idea world (still moving up in Figure 1.1) would be to use previously derived theorems or agreed-on assumptions, based on the selected model, in order to solve or simplify the mathematical model. For example, you might proceed to solve an equation by applying derived rules or axioms of equality. Finally, the problem solver returns to the real world (see the curved arrow moving down between the two worlds) to ensure that the idea-world answer (*e.g.*, the solution of an equation) is valid in the face of real-world constraints. Thus, the process depicted in Figure 1.1 provides another way of looking at the process schematized in the *Standards* on page 138.

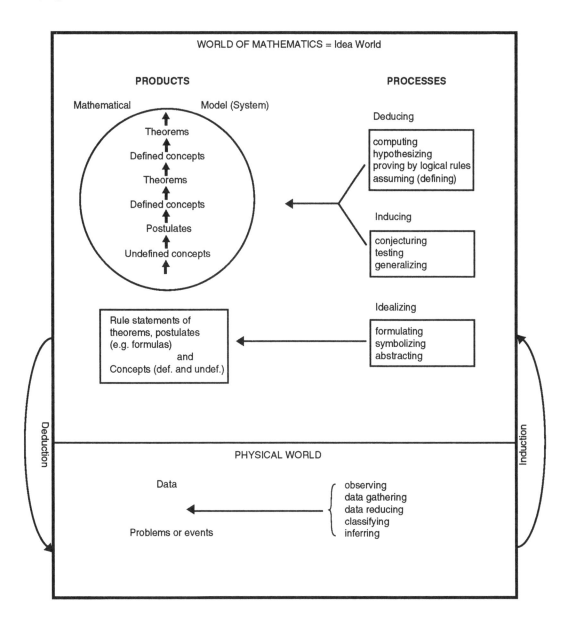

Figure 1.1: Model of Mathematics

The same process led historically to the mathematical systems we presently use and continues to be applied to invent new mathematical systems or to extend and adapt earlier ones. In the larger view, then, learning about mathematical models leads to learning about the nature of mathematics. In a narrower but very important sense, learning about and learning to apply mathematical models leads to improved problem-solving skills.

What Is a Mathematical Model?

So far we have given examples of mathematical models and alluded to the way in which they are selected. You have also been provided with a model and a schema that depict how one uses mathematical models to solve real-world problems. Now, let's examine a good working definition of a mathematical model, one that can be used with secondary students.

> A mathematical model refers to a set of mathematical terms and statements that appears to be an idealized, but faithful, reflection of data and/or events in the physical world. (Farrell and Farmer, 1988, page 79)

Mathematical terms and statements may include equations, graphs, or even a collection of points and lines. Notice the emphasis on the **apparent** one-to-one match between a particular real-world situation and a selected mathematical model. This is a characteristic of mathematical models that needs to be stressed with secondary school students. This can be done in the context of appropriate real-world problems. Here is one example.

Students are given the following information reported by a driver who negotiated the traffic in center city Boston on the way to the Interstate north to Maine. "It took me 20 minutes to go 7 miles, and that was just to leave the inner city. In the next 20 minutes, I was able to cover 15 miles, and then I got on Interstate 95 and drove 23 miles in 20 minutes." After the students compute the velocity in miles per hour for each stage of the journey, they should then graph the velocity against the time in hours. The time dimension on the graph can be scaled in quarter hours.

In the question-and-answer session that follows, the teacher has an opportunity to explore how the resulting step graph faithfully reflects one aspect of the trip (average velocity over each stage of the reported trip), while distorting another aspect (acceleration) of the real world of the trip. Students will generally immediately notice the discontinuities in the graph and recognize that the graph has the driver instantaneously accelerating to go from one step to the next. Yet, even with this mismatch between the mathematical model and some aspects of the real-world situation, the graphic model is useful in its representation of the traveler's image of the journey. It dramatically highlights the three stages of the journey and leads to a discussion about the reasons for the different stages. Teachers can also have students compare the given velocities with the legal maximum speeds for each probable portion of the trip. Thus, a values-oriented connection to social studies class can be added to the lesson.

(Teachers in other areas can develop similar travel descriptions by using highways and congested traffic sections known to their classes.)

Examples like the preceding one help students understand why the problem solver returns to the real world to validate the solution against real-world constraints. Secondary students need to be helped to see that validation is not the same as the checking process. In particular, validation is not used to identify errors in algebra but to identify points of non-correspondence between the model and the situation.

Consider the mathematical model that corresponds to the strength of a steel girder *(s)* in terms of its cross section *(x)*: $s = kx^2$. A graph of the function is a parabola with the origin as its turning point. However, half of the curve must be immediately disregarded in terms of the real-world situation, and other points will have to be rejected as unrealistic given limitations on the cross section of steel girders. It would be important to reassure your students that these unacceptable values are not reason enough to reject the model. The model still provides useful information to the engineer. The next issue should be a consideration of the points that are not being rejected. How are the values for cross section obtained? You'll need to help your students realize that when measurement (with the exception of counting) is involved, the result cannot be exact. Thus, the apparent precision obtained by finding the coordinates of the points on the parabola is just that—apparent. How do structural engineers deal with this fact—the potential differences between the mathematical results suggested by a model and the real-world situation?

This question dealing with the acceptable range of error can open up a profitable give-and-take on the role of mathematics in applications. It sounds as if this would be a way to teach toward outcomes listed under Standard 4: **Mathematical Connections.** It certainly would, and we'll revisit this question in Chapter 4 of this text. Once again, notice that an integrated approach helps you to teach in a way that achieves multiple standards. However, if we blurred the distinctions among the standards, some important student outcomes might not be emphasized.

Consider the questions in the following Activities.

Activities

1-19. Refer to the networks lesson in the Readings, page 140. How is the situation represented by a mathematical model? What are the points of correspondence and non-correspondence between the mathematical model and the actual situation?

1-20. In their study of insect noises, scientists have found that insect sounds are related to temperature. For the house cricket, for example, the following formula describes that relationship:

$T = 10 + N/4$, where *T* is Fahrenheit temperature and *N* represents the number of chirps per minute.

a. What are the probable points of correspondence and non-correspondence between this mathematical model (an equation) and the real-world situation? (Talk with a science teacher about the range of values one might expect.)

b. If secondary students were asked to use this functional relationship to draw a graph of cricket chirps versus temperature, how might the teacher help them to recognize the restrictions on the graph?

1-21. Seniors at Milatola High were given the opportunity to plan a rearrangement of furniture in the Senior Lounge. The custodians would actually move the furniture, but only if they were given a precise "map" of the desired location of each piece of furniture. The students decided to draw a scale diagram of the room on graph paper and then cut out scaled shapes for the piano, chairs, tables, hassocks, bookcases, *etc.* After considerable discussion and moving around of the shapes, consensus was reached and the shapes were taped in place.

a. Why might the custodians still have some trouble in moving from the physical model (itself based on mathematical models of scaling and geometry) to the real world of the lounge?

b. After all the furniture was finally in the lounge, the seniors realized that their model had not taken into account an important aspect of the furniture. What might have been ignored?

1-22. Refer to the table on page 00 in the Readings. Identify the mathematical models and their real-world counterparts for each of the everyday ideas in the table. Be sure to consider the appropriateness of the model for learning about the real-world phenomena.

As you probably have realized, the preceding set of Activities could be used with secondary school students who have studied the mathematics in question and are ready to consider the nature of a mathematical model. What are some other ways to teach toward this outcome? Some suggestions and activities are outlined in the next section.

Instructional strategies

First and foremost, in the simplest of cases, the teacher needs to help students realize how often mathematical models are applied. The middle school teacher might ask students to use pairs of the geometric shapes they have studied to design a tile pattern. Their results can be drawn on sheets of paper. They should be asked to discuss the advantages of the chosen shapes and the reasons why they did not select others. From such a class activity, the teacher can elicit the properties of the cut-out shape, differences between a cut-out shape and the mathematical model it represents, and other differences between the paper tiling and an actual tiling of vinyl. On the one hand, every time the teacher helps students to recognize that a particular physical model is not the same as the mathematical model it represents, the students are learning more about the nature of mathematics. On the other hand, when real-world choices are discussed, the students get some ideas about the use of mathematical models to solve real-world problems. Thus, every time a physical model is used, there is an opportunity to teach about the connection between it and the related mathematical model. Every time a mathematical model is used in conjunction with a real-world situation, there is an opportunity to teach students about the process of modeling.

However, because so often mathematics is learned as if there is a unique set of responses to every question, students have difficulty understanding that different meanings may be associated with the same mathematical model. In the era when number systems other than those based on ten were introduced, parents were astounded to have their children tell them that $1 + 1$ might equal 10, base 2. As far as they were concerned, when the symbols $1 + 1$ were written, there was exactly one correct response, 2. Of course, they were reading 10 as *ten* and were not aware that the digit 1 now represented 1×2^1. The fact that different meanings might be associated with the digits and the place values was completely beyond their experience.

The preceding example highlights an affective problem of many of our students. They may be accustomed to forging ahead to get the one correct answer and have little or no experience in thinking of mathematical symbols, equations, graphs, or shapes as tools that can be used to solve multiple problems. The fact that the same tool (*e.g.,* the same equation) might be used to solve problems from entirely different contexts might be outside their experience.

It is especially important that students understand this feature when considering mathematical models. Here is one way to approach that understanding. Design a homework assignment to practice the solution of word problems by algebraic methods. Include among the exercises at least two situations that have surface differences but will typically be translated into the same equation structure. In class, after any misconceptions or errors have been considered, ask the students to take a second look at the special exercises and to volunteer any patterns they notice. This encourages students to consider the fact that a particular mathematical model may represent more than one situation. Students can also be helped to realize that the same mathematical solution may take on different meanings when it is verbalized in terms of the different real-world situations.

Activities

1-23. Assume that you are working with algebra students on the possibility that the same equation may represent more than one real-world situation. Assume also that you have assigned the following exercise.

We have been writing equations as algebraic translations of word problems. The equations are actually mathematical models of the situation described in the word problem. In the following, you are given equations, but not the situations they model. Write a word problem that might be modeled by each equation. Each of these equations can represent any number of different situations. We'll share responses tomorrow and see what variety the class constructed.

a. Choose a unit on equations from a contemporary secondary school algebra text. Based on the content in that unit, write two equations that could be given to students with the instructions above.

b. For each equation you wrote in response to **a**, write two word problems that might be modeled by the equation.

c. Reread the last sentence in the instructions given to the students. What should the students expect will happen in the next class? Why is it important to include this sentence in the exercise?

1-24. Write a question similar to that in Activity 1-23, but provide the students with graphs, each of which is a mathematical model of more than one real-world situation. Again, ask the students to write questions or provide a description of a situation, that might be modeled by each graph. (Answer the question yourself in order to obtain sample responses to share with the students.)

1-25. Just as an equation or a graph may represent more than one real-world situation, so a geometric shape may be used as the mathematical model for more than one real-world situation. Secondary students and you see these examples daily but may not identify them as mathematical models of real-world phenomena. A square, for example, is the model used for floor and ceiling tiles as well as for some picture frames and windows.

a. Prepare a list of real-world phenomena that are modeled by a regular hexagon, a cube, a sphere, or a rectangular solid.

b. Devise an activity for secondary students to alert them to the multiple uses of some particular geometric shape as a mathematical model. Be sure to have them consider the rationale for using one shape over another in each case.

As these Activities show, teaching students to apply mathematical models in the process of solving problems is a gradual, developmental process. It begins with number sentences to represent simple situations, bar graphs and pictograms to picture data, geoboard diagrams to simulate a situation, and model-building to concretize three-dimensional structures. As the students move into middle and high school, the mathematical models become more sophisticated. Computer simulations and calculator graphs are added to the collection of potential modeling tools; and teachers have to help students become increasingly alert to the distinctions between the physical models, their mathematical counterparts, and the real-world phenomena being modeled.

The Vision: *Problem-Solving Disposition*

The authors of the *Standards* make it clear that attitude formation is important to the achievement of Standard 1, when they include the outcome : "...so that students can acquire confidence in using mathematics meaningfully" (NCTM, 1989, p. 75). They emphasize the importance of helping students become independent doers of mathematics. Calculators and computers are cited as tools to help students explore conjectures and verify results. As the NCTM authors note, confidence is built when students realize that their ideas and explanations are important to the teacher.

A Bridge to the Classroom

How do we help students gain confidence? Perhaps it would be helpful to reflect on the final words of the outcome: *using mathematics meaningfully.* Throughout this chapter, there have been references to choice of problem situations that make sense and are of potential interest to the students. There has also been emphasis on the importance of designing instructional strategies that correspond to the students' cognitive developmental levels and learning styles. The incorporation of all of these factors will help ensure that student learning is meaningful. The alternative—little learning or rote learning of mathematics—leads to minimal retention and negligible transfer. Under these bleak circumstances, there would be little chance that students could use mathematics **meaningfully**.

Now let's put the emphasis on the word *using* from the same phrase, *using mathematics meaningfully*. The students must have experiences in using mathematics—experiences with varying amounts of assistance from the teacher. There must be opportunities to interact with problem material, and there must be allowance for missteps in the process of trying to solve a problem. In an earlier section, you read that teachers need to communicate to students that problem solving does not flourish in an alarm-clock environment. Time—to think, to make mistakes, to reconsider steps, and to stumble and try again—is an essential ingredient in novel-problem solving. Confidence is gained when we have the time to succeed at something.

In the case of problem solving, the definition of success is broader than getting the right answer. Recall the project directions in which students are told to keep journals in which they will record all the strategies they try, how far each takes them, and what they try next and why. In this case, success is defined in terms of the choice of reasonable strategies—strategies that seem to hold promise.

Because the traditional mathematics assignment places a high premium on correct answers, process-oriented activities need to be developed carefully. Teachers have to reveal their own fallibility when faced with novel questions. One convincing way of doing this is to take the risk of attempting a problem, new to you, the teacher, with class members as helpers. Sharing thinking processes and even openly admitting to a temporary stalemate are powerful ways of modeling the attitude of a problem solver. (For more information on the role of the teacher as a human model and the need for a supportive classroom milieu, see pages 141 and 142 in *Secondary Mathematics Instruction*.)

Don't forget the importance of the student's peers in developing or restricting the development of confidence. Here is one instance in which cooperative learning groups can provide support through teamwork to a student struggling with problem-solving skills.

Confidence, therefore, is developed over time. As you might expect, a confidence-building classroom milieu is equally important in helping students achieve each of the other standards outlined in this text. As you interact with the reading and activities in those chapters, take the time to reflect on the kind of feedback-getting and -giving strategies needed to promote a positive disposition towards mathematics.

Activity

1-26. Parodies of the so-called *New Math* era sometimes include songs that claimed:

> *"The important thing is to understand what you do, rather than to get the right answer."* (A line from Tom Lehrer's song, *New Math*)

This claim is definitely not what the proponents of *New Math* intended, nor is it what the authors of the *Standards* intend. However, some parents might believe that is the kind of attitude an open-ended assignment will foster. How could you dispel this belief? What might you do, as a teacher, to be sure that students don't acquire this attitude?

References

Barrett, G. B.; Bartkovich, K. G.; Compton, H. L.; Davis, S.; Doyle, D.; Goebel, J. A.; Gould, L.D.; Graves, J. L.; Lutz, J.; and Teague, D. J. 1992. *Contemporary precalculus through applications.* Dedham, MA: Janson Publications.

Davidson, N., ed. 1990. *Cooperative learning in mathematics: A handbook for teachers.* Menlo Park, CA: Addison-Wesley.

Farrell, M.A., and Farmer, W.A. 1988. *Secondary mathematics instruction: An integrated approach.* Dedham, MA: Janson Publications.

National Council of Teachers of Mathematics. 1989. *Curriculum and evaluation standards for school mathematics.* Reston, VA: NCTM.

Curriculum Standards for School Mathematics

STANDARD 2: Mathematics as Communication

In grades 5–8, the study of mathematics should include opportunities to communicate so that students can

— *model situations using oral, written, concrete, graphical, and algebraic methods;*

— *reflect on and clarify their own thinking about mathematical ideas and situations;*

— *develop common understandings of mathematical ideas, including the role of definitions;*

— *use the skills of reading, listening, and viewing to interpret and evaluate mathematical ideas;*

— *discuss mathematical ideas and make conjectures and convincing arguments;*

— *appreciate the value of mathematical notation and its role in the development of mathematical ideas.*

In grades 9–12, the mathematics curriculum should include the continued development of language and symbolism to communicate mathematical ideas so that all students can

— *reflect upon and clarify their thinking about mathematical ideas and relationships;*

— *formulate mathematical definitions and express generalizations discovered through investigations;*

— *express mathematical ideas orally and in writing;*

— *read written presentations of mathematics with understanding;*

— *ask clarifying and extending questions related to mathematics they have read or heard about;*

— *appreciate the economy, power, and elegance of mathematical notation and its role in the development of mathematical ideas.*

· Reprinted with permission of the publisher. See acknowledgements.

Chapter Two

Mathematics as Communication

Communication? What do the authors of the *Standards* mean by that? Mathematical expressions are often collections of letters and numbers, and mathematical sentences are usually expressed as equations or inequalities. In fact, there are those who define mathematics as a language and suggest that for students to learn it, one just needs to teach them this new way of representing ideas. Is that what the authors of the *Standards* had in mind?

Study the outcomes under each of the grades 5–8 and grades 9–12 versions of the second standard listed on the opposite page and formulate examples of some kinds of communication referred to by the NCTM authors.

The Vision: *We Need Communication!*

According to the popular press, school guidance counselors, and a number of talk show hosts, communication holds the key to better parent/child relationships, to longer lasting marriages, and to prevention of ethnic disturbances. Yet, two antagonists who claim the other does not communicate are likely to concede that they talk, but they still don't communicate. What's missing? Sometimes the listener endows the words with a different meaning than is intended by the speaker. Sometimes there isn't any real attempt to listen and respond to what has been said. These everyday occurrences of communication breakdown provide clues to the kind of communication needed to improve the learning of mathematics.

First, as the authors of the *Standards* noted, the special language of mathematics allows a number of diverse and complex real-world phenomena to be represented succinctly. Symbols and symbolic statements such as π and $A = (1/2)bh$ represent concepts or procedures, respectively; they are not themselves the concepts or procedures. However, the authors emphasized that without frequent and explicit consideration of the relationships between the symbols and the concepts or procedures they represent, the students are liable to view the symbols as objects to memorize (NCTM, 1989, p. 78). Thus, students may be hearing mathematical language, but learning something quite different from what the teacher intends. The communication loop has broken down.

If the teacher ignores this breakdown and blithely continues to use the language of mathematics to teach *new* concepts and procedures, the student may listen, but only on a surface level. Imagine being in an architecture class taught in a language foreign to you and trying to identify clues of any kind to decide what you're asked to do.

> *Some listeners have characterized mathematics lessons as lessons in a 'foreign' language. Imagine the plight of students not proficient in this language if their teachers were to conduct lessons almost entirely in mathematical language.* (Farrell & Farmer, 1988, p. 167)

Thus, the second breakdown occurs when the student stops trying to listen in a meaningful way, or to connect learned ideas and images with the new ideas being taught, and seizes on surface features that seem to be necessary for success on a day-to-day basis.

You probably have met college students and adults who consider mathematics a mystery and its language incomprehensible. There may even have been times when you found yourself groping for meaning in new groups of symbols being presented without reference to prior concrete understandings. At those times, you were not ready to learn the concepts presented at an abstract level. In the language of cognitive development, you were not yet able to operate at a formal operational level in that particular area.

Activities

2-1. Refer to Ms. Blumenstalk's approach to area proofs on page 142 in the Reader.

a. Identify the ways in which she incorporated concrete referents for the various concepts and procedures related to area.

b. Identify the points in the sequence of lessons in which the students used particular communication skills to clarify their understanding.

c. Some of Ms. Blumenstalk's tenth graders may have been able to function on a formal operational level. How would these hands-on, highly concrete lessons have helped them; or would such lessons be wasted on them? (For further background information related to this question, see the section on the *nested stages* on pages 54–57 in *Secondary Mathematics Instruction*.)

2-2. Differences in the way students process information affect the learning of symbol/referent relationships. Suppose that one of your algebra students is a skilled symbol-manipulator and correctly solves systems of equations. Devise a question or activity to get feedback on that student's understanding of the procedures and the effects of those procedures. (For background material, refer to the section on *Processing Information* on page 64 in *Secondary Mathematics Instruction*.)

Another aspect of the communication problem is found in the insistence by some teachers on precision of language and on reproduction of exactly the same definition statement given in a text or in notes. Demands of this type, when made before the concepts and principles have been understood, lead to the view that mathematics is a subject that allows for no creativity and no variation in approach. These negative views of mathematics are not likely to result in the problem-solving disposition referred to at the end of the first chapter of this text. Is there a time for insistence on precise statements of definitions and use of accepted symbolic representations of concepts? Yes, definitely, for we **do** want our students to be able to communicate with other mathematics students and to understand the power of this universal language. However, that insistence must be delayed until students are thoroughly grounded in the ideas being represented by the language.

Speaking, listening, watching, reading, writing—each of these five communication skills is mentioned in one or more of the outcomes under Standard 2. Speaking and writing are familiar ways to express our mathematical learnings, aren't they? All of us have been asked to state answers to problems, recite definitions, or perhaps, explain why we chose to use a certain strategy. If these have been our only experiences with the skills of speaking and writing in mathematics class, we probably associate writing with quizzes, tests, and homework assignments and speaking with an intermediate evaluation step. However, the authors of the *Standards* had a much different image in mind when they used these words in their outcome statements. They extended the range of appropriate speaking and writing opportunities, both in terms of context and audience. For example, they recommend that students express a recently learned rule in their own words, rather than in the words of the text. Small groups of students can share their diverse written statements and discuss similarities and differences. They can also evaluate the statements to identify those that would be easier for a younger child to understand or those that contain no extraneous details. Thus, the students would be communicating with each other in an attempt to understand a mathematical procedure.

It is important to study each of the outcomes under Standard 2 with this broader view of audience and context in mind. Although this view does not exclude the teacher as audience or more typical written and spoken exercises, it does relegate them to another time and a different purpose.

Did you notice that in the example of small group sharing, listening would have to be an important part of the interaction? Would either of the two remaining skills, watching or reading, have been used in the small groups? If you decided that reading each other's written responses might have occurred, you're right on target. Notice the difference in the outcomes listing the reading skill on the 5–8 and 9–12 grade levels. If it isn't apparent, return to the descriptions that follow the list of outcomes in the *Standards*. Notice that the expectation for the 9–12 students is that they will be reading mathematics **text** material. This may not seem unusual to you. After all, you undoubtedly are expected to read material in your calculus, probability, or foundations of geometry texts. However, pre-college students, even in advanced classes, seldom have to read the text.

The material is usually digested for them and presented to them by the teacher.

It seems that the only skill not needed in the small group example is watching or viewing. Once again, the authors of the *Standards* surely had more in mind than having students watch the teacher or another student write an example on the board. For example, the students might be expected to watch the teacher demonstrate transformations of a rhombus with fixed sides by flexing a georule or carpenter's ruler, or with graphics or geometric technology. The students would be asked to make a conjecture about the relationships of area and perimeter as the figure changes.

Activities

2-3. In both the grades 5–8 and 9–12 sets of outcomes, one outcome begins with the words *reflect* and *clarify*.

a. Study the two outcomes, but be sure to read them in reference to the over-arching statement that begins the particular set of outcomes. What differences, if any, are expected at the two sets of grade levels?

b. Refer again to the parabola activity on pages 129-130 of the Reader. In that activity, the participant is asked to *reflect* on his or her own thinking. Subsequently, an interpretation of what occurs when one *reflects* is given. What aspects of this activity are likely to enhance reflection? Which communication skills are needed during this activity?

2-4. In both the grades 5–8 and 9–12 sets of outcomes, one outcome begins with the word *appreciate*.

a. Study the two outcomes and again identify any differences in expectations at the two sets of grade levels.

b. Refer to the suggested approach to the modeling of common algebraic identities found on pages 143-144 of the Reader. Using strips and square shapes, model

$a(b + c) = ab + ac$ and $a^2 - b^2 = (a + b)(a - b)$.

Compare your efforts with those of a colleague. Consider how difficult the process would be if, as was the case for the Greeks, you did not already have a result to work toward.

c. The modeling examples from **b** could be demonstrated by the teacher with a resultant saving of time. What might be lost with this approach to the lesson if one of the objectives was to have the students appreciate the powerful role played by the invention of symbols? What communication skills might help the students reach this objective?

It should be clear that these communication skills are not learned in an incidental fashion. They must be the objects of instruction at all levels. For instance, even when the students are in an advanced mathematics class, you can't assume that they read mathematics text material in appropriate ways. Moreover, the observation that all of the students are quiet and are gazing toward a demonstration table doesn't necessarily mean they are viewing the demonstration correctly. In the next section of this chapter, you will find examples of strategies designed to help students learn and use communication skills.

A Bridge to the Classroom

A careful reading of the beginning of Standard 2, as written for grades 5–8, makes it clear that the **student** is at center stage. The teacher must provide opportunities for the **students** to communicate. In order to provide time for such opportunities, there must be a corresponding reduction in the amount of talk by the teacher. This rather significant change in traditional classroom interaction makes it imperative that the communication activities be designed with care and that attention be given to developing student communication skills.

Two important aspects of speaking and listening emphasized in the *Standards* are (1) that students interact with students as well as with teachers, and (2) that students communicate mathematical ideas in ways other than the final, polished, symbolic form. Thus, student pairs, small groups, or assignments that require students to share ideas are some of the possible modes that a teacher might use to encourage speaking and listening. Some organizational suggestions for effectively implementing these modes were considered in Chapter 1 of this text. In this chapter, we'll consider the communication processes themselves in more detail.

First and foremost, students have to be made aware that the teacher is interested in the kind and quality of oral and written comments, questions, and explanations they use with each other. Students will be convinced that this is important if the teacher gives feedback on the conversations they have within a group, takes the time to have other students listen to a student comment or question, and helps students to explore another relevant avenue without simply ending discussions abruptly.

Activities

2-5. Refer to the probability laboratory lesson found in the Reader on page 145-146.

a. How did the teacher get feedback on the quality of communication in the groups?

b. When Saul and Alice disagreed on the procedure for getting a probability for two events occurring at the same time, what specific actions did the teacher take to encourage them to communicate further? How did the teacher reinforce the importance of student-to-student communication with the entire class?

2-6. Refer to the Primes lesson plan on pages 132-136 of the Reader. Answer the following after reviewing the section of the plan where the teacher elicits patterns from the students.

a. What is the "checker" definition referred to in this section?

b. Refer to the explanatory material at the end of the Primes plan, in which Mrs. Lopez's view on the acceptability of this definition is presented. What is her view and why?

As the teacher in the Primes lesson emphasized, the importance of having students state their own definitions cannot be understated. She asserted that students are more likely to understand and retain definitions which they construct themselves on the basis of a concrete activity. How might improved understanding be explained in terms of cognitive and social interaction or the nested stages model? How might improved retention be explained? (Information on related background research and theory can be found in *Secondary Mathematics Instruction*. See the sections on pages 54–59 for the nested stages and interaction, pages 148–149 for retention, and page 136 for concept-learning.)

  In both of the sample lessons referred to in Activities 2-5 and 2-6, student pairs or small groups were the organizational stimulus for student interaction. Yet, veteran mathematics teachers often avoid such grouping arrangements because they find the changed atmosphere a threat to control and are not comfortable with, or not sure about, their own new roles. Beginning teachers who believe in the potential value of the communication among students are apt to ignore noisy groups if the students appear to be talking about the mathematics in question. Unfortunately, one noisy group seems to lead to others who become louder in turn, just to be heard. In the worst case scenario, students who can no longer hear each other turn to other activities. Not only has the effectiveness of the lesson been diminished, but the teacher has unwittingly provided a negative pattern for small group behavior in the class. Study the ways in which the teacher in the probability lab lesson handled loud voices (Reader, pages 145-146).

> Hint: It's important to provide students with some reasonable guidelines when they work in small groups or in pairs. In particular, the teacher has to emphasize how important it is that each person in the group be able to hear the others talk. This suggests that they shouldn't be distracted by loud voices from nearby groups and shouldn't have to raise their own voices to be heard. The teacher can teach the students to respond to nonverbal signals to keep the groups on target. For example, the quick flicking off and on of the light switch can signal the need for immediate silence so that a student idea can be shared, a demonstration repeated, or further instructions given. Another effective nonverbal signal, used by the teacher in the probability lab example, is the finger to the lips to indicate to a particular group of students that they need to quiet down. It goes without saying that the teacher shouldn't needlessly interrupt the groups by talking to one student or one group in a loud voice. (See pages 40–42 in *Secondary Mathematics Instruction* for more examples of giving feedback.)

Activities

2-7. Listening, watching, and reading were also used by the students in the probability lab lesson (Reader, pages 145-146).

a. What are the specific circumstances in which each of these skills were needed?

b. In which cases was there a breakdown in these skills? How did the teacher react to the breakdown in each case?

The less interactive skills of listening, watching, reading, and writing need attention from every subject-matter teacher. How often have you heard the complaint: "They just don't listen to the directions !" There are at least two reasons for this apparent heedlessness. First, the students may have learned that the teacher always repeats directions if asked. Thus, it just isn't important to listen the first time. Second, the students may not have learned to listen for emphasis in voice, summarizing cues, and the like. Similar reasons could be given for poor watching and reading skills, along with the possibility that these may be skills seldom used in mathematics class.

> Hint: Tape recordings are relatively inexpensive ways of standardizing information to be given to students. If a repeat of oral information is desired, the playback feature ensures an exact repetition. The teacher might tape a set of instructions that students are to follow in order to complete examples, draw sketches, or perform lab work. (For other uses of audiotape recordings and for suggestions on the kind of teacher preparation needed, see page 304 in *Secondary Mathematics Instruction*. For additional suggestions on the use of equipment, the preparation of the students, and the integration of audio-visual and technological activities into lessons, see pages 20 and 21 of *Secondary Mathematics Instruction*.)

Activities

2-8. Answer **a** and **b** on the basis of the transcript in Sample # 1 and **c** and **d** on the basis of the transcript in Sample # 2.

Sample 1: *You need a sheet of graph paper, a sheet of paper for notes, a straight edge, and a pencil.* pause *You will be asked to draw some figures on the sheet of graph paper. Each instruction will be repeated, but only once. I recommend that you listen the first time and make any notes you think important. The second time, draw the figure.* pause *Ready. Here we go. Draw an isosceles triangle, with a base of 6 units and a leg of 5 units.* pause pause *Repeat: Draw an isosceles triangle, with a base of 6 units and a leg of 5 units.*

a. If the teacher tours the room systematically as the students are drawing, what feedback might reflect good or poor listening skills?

b. How might a student show evidence of good listening skills, but still not be able to complete the drawing correctly?

Sample 2: *You'll be listening to student explanations of various concepts. Some of the explanations contain subtle errors; some are based on interesting conjectures worth exploring. You will be playing the role of the teacher, who must evaluate the worth of a student oral response as it is given. However in your case, each explanation will be repeated, just once. Listen carefully; take notes on important statements. Then work within your small groups on your response to the student explanation.* pause pause **Myron, how would you explain the concept of function, in your own words?** pause pause **I'd describe it as a type of dependency relationship, in which changes in one variable cause changes in another variable.** pause pause *Now you will hear a repetition of the teacher's question and Myron's response.* pause **Myron, how would you explain the concept of function, in your own words?** pause pause **I'd describe it as a type of dependency relationship, in which changes in one variable cause changes in another variable.** pause pause *Now, you may begin to discuss the quality of Myron's response in your groups. Please keep a record of group ideas. I'll signal when it's time to listen to another student explanation.*

c. What feedback will signal poor or good listening skills as the teacher systematically tours from group to group?

d. What mathematical understandings might the teacher expect to elicit through group work on Myron's response?

As these activities show, developing listening skills does not need to be done at the expense of learning mathematics. Strategies may be designed so that both are the objects of the lesson. In order to further identify student problems in listening and emphasize its importance as a skill, the teacher might add a question to the activities outlined in Activity 2-8, part a. The students might be asked to write down (1) anything they think they missed while listening, (2) clarifying questions they would have asked, if allowed to do so, or (3) any difficulties they had in listening to the tape. If the teacher collects the graph paper with the drawings and this information on the reverse side, a more systematic identification of both listening problems and mathematical problems is possible.

Isn't the collection of this kind of feedback a type of evaluation? Yes, it certainly is. If we examine effective developmental lessons, we will always find elements of this kind of evaluation. The teacher can decide whether the instructional strategies or the pace of the lesson need to be modified. When you read Chapter 5: Evaluation Standards, you will find more examples illustrating this view of evaluation as part of, not separate from, instruction.

Watching or viewing skills need to be honed just as carefully as listening skills. Students must be told exactly where to look, what kinds of information they are to observe, and whether they will be expected to report orally or in writing on the data. It is sometimes useful to perform a demonstration twice, with the intention of focusing students' attention on specifics the second time.

For example, suppose the teacher planned to stretch and shrink shapes on an overhead geoboard or using technology and a monitor the whole class can see. The first time, the teacher might just give the general direction: *Watch what I'm doing with the shapes on this geoboard.* Before repeating the demonstration, the teacher might say something specific like: *Pay particular attention to what changes and what stays the same. Watch sides, angles, overall shape. Don't say anything out loud. We want everybody to have a chance to think. I'll ask for hands when I've finished with the demonstration.* In the subsequent question-and-answer session, students get a chance to hear a variety of observations.

The teacher will need to take some straw polls to ensure that all observed the same details. If there are students who hadn't noticed some aspect of the demonstration or if there is disagreement about the reported observations, the teacher can repeat the demonstration until all are satisfied. (For more information on demonstrations, see pages 15–16 and 297–298 in *Secondary Mathematics Instruction.*)

Activities

2-9. Read about the Bicycle demonstration on pages 147-148 in the Reader. Suggest some specific questions or statements the teacher might use so that the students would know what to look for during the demonstration.

2-10. Suggest subjects for photos or slides that could be used to help students understand a mathematical concept or principle. What would the students be asked to look for in the series of photos or slides?

As noted earlier, the reading of mathematics material needs to be encouraged as early as possible and extended to substantive mathematics text material at the 9–12 grade levels. At the 5–8 grade levels, students might begin with Idea sheets copied from NCTM Newsletters or journals. These Idea sheets include some expository material, directions, and spaces for student responses. Thus, the same material can provide practice in both reading and writing skills. With the exception of children whose reading ability is far below grade level, the expectation should be that these materials can be understood by the children of the designated grade levels. However, the students need to be taught to read this material differently from fiction. The teacher may want to place a time limit on the silent reading of a short section of the material and then have students summarize the ideas presented or the central question.

Notice that **silent** reading is suggested at this stage. The silence must be enforced and must extend to the teacher. The teacher must resist the temptation to read the passage aloud or students will learn that they really aren't expected to read for understanding.

If the material requires English sentences to be translated into mathematical sentences, students must pay particular attention to the importance of every word and phrase, and especially to the mathematical meaning of words such as *and*.

It is also important to recognize the difficulty posed by using certain symbols in both a mathematical and a standard-language context. One of these difficulties was pointed out by Farrell (1969), who found that young adults who had been educationally disadvantaged seemed to be blocked when faced with the use of a letter to stand for a variable. If the variable was represented by a frame (□), and the students were instructed to fill in the frame with the replacement that would make the sentence true, they were able to make sense of the equations. (For background information and instructional suggestions related to other misconceptions in algebra, see pages 60–61 in *Secondary Mathematics Instruction*. For background information on some of the other special problems of translations between everyday language and mathematical language, see pages 174–175, and for information on levels of translation, see page 178 in the same source.)

In contrast to these suggestions on practice, there are times when reading mathematics aloud is a waste of time and may interfere with the meaningful development of that skill. When might these occurrences be? Having a student read aloud every question that was on a test or assigned for homework leads to boredom in the classroom. If you've ever listened to these droning repetitions of questions which all the students have already read and worked on, you already understand why the practice might be detrimental to the development of reading skills.

Other materials suitable for developing reading skills are short essays about mathematics, such as those in the *Readings in Mathematics* series (Adler, 1972). Two or three questions to help the students focus on the material can be provided along with the reading assignment.

One caveat is in order. Consideration of the reading and interpretation of graphic material has been omitted here, not because these skills are unimportant, but rather because they have been accepted as important and, thus, are already a standard part of the curriculum.

Activities

2-11. For this activity all students must have the same kind of basic, four-function calculator. They should know how to use their calculators and how to interpret and execute keystroke sequences. They should be provided with the following handout and told to work in pairs to complete the handout. After reading the student handout, respond to **a** and **b**.

Calculator Patterns

Enter each of the following numbers (5, 8, 11, 14, 27, 103) for a in the keystroke sequence that follows:

a $\boxed{+}$ 2 $\boxed{\times}$ 4 $\boxed{-}$ 4 $\boxed{\div}$ 4 $\boxed{-}$ a $\boxed{=}$

As you complete the calculator activity, record your results in the table below. (Notice that the first result has been written in so that you can check your work.)

Input	5	8	11	14	27	103
Result	1					

What is the pattern? Try to explain why it happens. Write your ideas on the back of this page.

a. Work through this calculator exercise and identify the potential reading problem(s) for students.

b. What is the advantage in having student pairs work on this activity?

If you thought that students might use different numbers for the first and last *a* in the keystroke sequence, you've identified a common student error. Did you respond that each student in a pair can serve as a check on the other? They can also help each other refine writing skills and clarify mathematical ideas. Remember, this exercise is intended to help develop reading skills, and not to be a final assessment of those skills.

When teachers use the calculator pattern sheet shown in Activity 2-11, they are helping to develop their students' writing and reading skills. What other examples of the use of writing have been alluded to in this section? Take the time to go back and review the section. You should find examples in small group work, in response to demonstrations and to the tape recorder lessons, and in conjunction with reading-comprehension tasks. It is almost a *sine qua non* that writing should occur in small group work. Whether the students are engaged in a hands-on activity or a proof evaluation, there are many benefits in having all students write (1) conclusions, (2) major ideas or concerns, (3) records of data obtained, and (4) questions for the teacher.

> Hint: When initiating small group work it helps to design a handout with specific questions and response spaces. Even though the teacher may direct students to work toward group consensus, which will be reported in oral form to the whole class, individual sheets should still be collected. This confirms the importance of individual ideas and holds each student, as well as the group, accountable for the work. According to Davidson (1990), one criterion associated with success in the use of small groups is the incorporation of both individual- and group-accountability mechanisms.

Writing in mathematics class has increasingly been recognized as a factor influencing students' learning. It can also be a powerful diagnostic tool for the teacher. Here are a few suggestions on ways to incorporate writing into your classroom.

The first example has been used by several junior high teachers. Just before a unit on ratio (or fractions or symmetry or other), the teachers ask the students to write a few sentences explaining ratio (or fractions or symmetry or other), as they might to students who have never heard the term before. The writers are told not to put their names on the papers since the teacher is primarily interested in class patterns and ideas. This is an in-class exercise, probably given in the last 5–10 minutes of a class. The teacher collects the papers and looks for common misconceptions, as well as major strengths.

This is a powerful diagnostic technique that students will take seriously if the teacher follows up with general feedback and indicates the usefulness of the written ideas to help plan teaching strategies. Students can begin to realize that error patterns are not associated with carelessness, laziness, or stupidity, but with the essential human trait of trying to make sense out of things, to make complicated things simple, and to find patterns.

Diagnostic exercises such as this are worth using at intervals during the course. I can give you no guidelines as to when there might be too much of a good thing. In all the classroom teacher testimonies I've read or heard of, repetition seems to grow more useful as long as the students sense that the exercise is meaningful and is being incorporated into the teacher's planning. In this first example, the students communicate with the teacher, but necessarily get no individual feedback on their writing. The question posed is a broad one and it is expected that responses may be wide of the mark.

The second example differs from the first, particularly in the amount of time both student and teacher must give to the task. In a twelfth grade mathematics class, the following take-home assignment was given:

> Write two or three paragraphs beginning with the topic sentence :
>
> *I dreamed I was one of three zeroes of a polynomial.*
>
> Be sure to use related mathematical concepts and/or processes in valid relationships to one another.

In this example, the students have wide latitude as to the depth and creativity of the final result. The teacher is interested in individual results and promises to share, with permission, some of the papers with the entire class. Although creativity should be praised with written comments, students should not be penalized for failing to write in a creative fashion. (See the suggestions on page 209 in *Secondary Mathematics Instruction* for ideas on the kind of feedback the teacher needs to provide to the students.)

The third suggestion is based on concept mapping. The teacher may write a list of concept labels from the most recent unit on the board or on an overhead acetate. Pairs of students are then asked to draw concept maps, discuss the maps, and be prepared to share their results with those of their peers. Some teachers call these *spider graphs* or *webbing*. (For background on uses of this process with students, refer to pages 143–144 of *Secondary Mathematics Instruction*.)

A variation on this third example is a take-home assignment in which each student is asked to write two questions relating pairs of concepts from the list. In the next class, small groups can be asked to share their questions and come to a consensus on the two most important questions for the whole class to consider.

If the teacher's purpose is to use writing as a way to get students thinking about **Mathematical Connections**, an assignment such as the following might be given to students in grades 5–8.

> *Write a few sentences about the mathematics used in your science class. Is there a difference in the way the science and mathematics teachers approach the same mathematics? If so, what are some of these differences?*

The next two examples were designed to help students make a bridge from mathematical language to ordinary language and back. In these exercises students are asked to write in ordinary language what some mathematical sentence or diagram might depict. Here are two examples. The first was used at the senior high school level; the second, at the middle school level.

(1) *Describe, in writing, a situation that this graph might be picturing.* (Any type of graph previously studied may be used.)

(2) Write *a word (story) problem that could be translated into this equation (number sentence).* (Use an equation-type or number sentence with which the students are familiar.)

Also see the Reader (page 138) for an example of a long-term assignment intended to provide students with experiences in reading, researching and formal writing. If this is the first time the students have been assigned a mathematics research paper, the teacher needs to help the students get started.

> Hint: Provide the students with a starter list of references and with opportunities to share early versions of the paper with you. In some schools, teachers of several subjects cooperate on this kind of assignment. Each teacher designates the particular objectives he or she will assess, and the student hands in a copy of the paper to each teacher.

Students may also be given a long-term assignment to write a poem, a play, or a story using mathematical concepts and/or principles from a unit they have studied. As in the case of the complete-the-sentence example, the premium is on the validity of the mathematics. One warning! Some students fear demands on creativity, despite the teacher's assurances. Thus, you might want to allow a choice between this assignment and some other long-term assignment, such as a geometric design project, a library report, a computer project, or a calculator project. An example of an eleventh grade student's essay, in response to one of these long-term assignments, is provided here.

Christmas Rush

From the center of the city for nearly a radius of two miles, masses of humanity swelled into it, almost disproving the property of impenetrability. At the intersections the two long straight lines of automobiles were perpendicular to each other and others almost coincided. The projection of the cacophony was tremendous. In a busy department store, children with large, beaming orbs gazed at a toy train making an elliptical run on a rectangular table. Gaily wrapped containers in the forms of cylinders, cubes, polygons, rectangles, triangles, and squares could be seen everywhere. From the midpoint of the crowd, the focus was placed on a large, red-suited ponderous man with an enormous circumference, and a long white beard sloping from his chin. His knee reciprocated to the bouncing hypotenuses of exuberant children relating infinite wishes. And so another Christmas comes to this old, whirling sphere moving along its oval orbit on a tilt of 23 1/2 degrees from the perpendicular.

The final example can be initiated at any grade level, including college, and requires students to keep a journal or log, daily or weekly, with writing done in class. They can respond to questions such as: *What have I learned? How do I feel about it? What don't I understand?* There are many variations on this example. Some teachers simply hand out sheets with one or more of these questions on them at the end of every unit or perhaps at the end of a week. As in the case of so many of the suggestions in this chapter, students will take the time to write their perceptions, rather than what they think the teacher wants to read, under certain circumstances. They have to be sure that the comments will be read and that an attempt will be made to deal with problems. They soon learn whether the teacher believes that the journal writing is important by the way with which it is dealt.

> Hint: One way to emphasize its importance is for the teacher to write responses to analogous questions while the students are writing. It is helpful to do this when students are just beginning logs and the teacher is building their confidence by providing in-class writing times. The teacher might share the written result on an overhead acetate. In this way the students are able to see the unpolished version of the teacher's perceptions. It's important that the teacher play fair and write spontaneously.

There are many articles on the subject of writing in mathematics classes. Three particularly helpful ones are those by Burton (1985), Davison and Pearce (1988), and LeGere (1991).

Activity

2-12. Choose any two of the writing examples provided in this section, and study the outcomes under the appropriate grade level range for Standard 2. Which of these outcomes might have been attained by students who completed either writing example?

The Vision: *The Nature of Mathematics*

In the preceding section, there have been repeated references to the need to design strategies that help students to develop skills and also help to promote mathematical understanding. What clues in the outcomes under Standard 2 refer to promoting mathematical understanding? You most likely identified phrases, such as *model situations, convincing arguments, common understandings.* However, these phrases represent only a general level of understanding. Are there phrases that suggest specific aspects of mathematical structure that students should study?

Study the third outcome in the grades 5–8 set. In this outcome, the authors highlight the role of definitions. The word *role* is particularly important here. Students, according to the authors of the *Standards*, need to develop an understanding of the role of definitions in mathematics. This is a very specific reference to the nature of mathematics. In the fifth outcome in this grade range, the processes of *conjecturing* and *making convincing arguments* are identified as ultimate goals of mathematical communication. Both of these processes are intrinsically related to the structure of mathematics.

Did you select the first outcome, because it refers to the ability to *model situations* ? I would agree that a consideration of mathematical models is also specifically related to developing a student's understanding of the nature of mathematics. A section was devoted to this concept in Chapter 1. At this point, it would be helpful to re-examine that section with particular attention to the communication skills being used in the work on mathematical models.

How do the grades 9–12 outcomes build on the grades 5–8 outcomes, with regard to the nature of mathematics? Again the role of mathematical definitions is emphasized but in a different way. Now students are expected to be able to use mathematical language and symbolism to *formulate definitions and express generalizations.* An examination of the descriptive material following this set of outcomes in the *Standards* shows that the authors expect these students to understand the axiomatic nature of mathematics, to give more sophisticated arguments and, in some cases, to write formal proofs. Understanding, and skill in developing, proofs are given in-depth attention in Standard 3, the subject of Chapter 3.

A Bridge to the Classroom

As noted earlier, the model depicting the structure of mathematics (Figure 1.1) was explored in Chapter 1. In this chapter, attention will be given to just one aspect of mathematical structure—the role of definitions.

The Role of Definitions

An examination of Figure 1.1 shows that the phrase *defined concepts* appears twice in the circle that is designated as a *Mathematical Model (System)*. There is also an upward-directed arrow at the top of all the phrases in the circle. This is a reminder to the reader that when a mathematical system is originally developed, mathematicians make a few definitions and attempt to derive as much as possible using these definitions, the undefined concepts, and the postulates they have written. When mathematicians desire to further develop the system in order to derive more results, they construct additional definitions. The final upward-directed arrow is suggestive of the on-going nature of this process, in which additional definitions might be invented at later stages of the process.

Since most secondary textbooks include only those mathematical systems that are thought to be in final form, the authors often provide all relevant definitions at the beginning of a chapter. Thus, students are unaware of the dynamic interplay that has occurred and still occurs in the development of mathematics. The definition statements, in final form, often seem so obvious that students are not introduced to the trial-and-error process that preceded these final statements. Thus, they are not made aware of the human aspect of mathematics.

Teachers can help students appreciate the tentative nature of these early definitions and the importance of the definition in deriving generalizations. Here are several suggestions: (1) have students read and discuss some of the controversies in the history of mathematics; (2) let students construct their own definition of some concept or let them consider alternate definitions of a concept; (3) provide students with examples of the ways in which definitions have been extended by mathematicians. The activities that follow are designed to help you reflect on this aspect of mathematical structure.

Activities

2-13. Read the Kite lesson ideas on page 137 of the Reader.

a. In this lesson, students are given a definition of a kite and are asked to make up their own definitions and perhaps their own labels for later definitions. Why might it be helpful to begin by providing the students with a starter definition? Suggest some additional definitions that might be useful in this lesson.

b. With a colleague, talk about the way(s) in which this lesson might help secondary students to appreciate the role of mathematical definitions.

2-14.a. Derive the rule for division of a^m by a^n, with $m > n$ and m and n being natural numbers.

b. Read how Mr. Greenberg explained the definition of a^0 on pages 149 of the Reader. How did he confuse *defining* with *deriving*?

c. How could you explain the meaning of a negative exponent? Must this also be defined, rather than derived?

d. How could examples like these help students to realize that definitions are not capricious statements, but thoughtful inventions by mathematicians?

2-15. How did Euclid attempt to define all basic terms? Why was this attempt doomed to failure?

2-16. Refer to Mr. Ramirez's lesson in Introductory Algebra on pages 150-152 of the Reader.

a. What rationale does Mr. Ramirez present for constructing the new definitions?

b. How does this approach illustrate the ways in which mathematicians extend early definitions in order to encompass a greater variety of cases?

2-17. Identify some other examples in mathematics in which students typically are presented with a subsample to which certain rules apply and later may have to alter definitions in order to extend the rules to the larger sample.

Throughout this chapter, there have been repeated references to aspects of **Mathematical Reasoning**, the subject of Chapter 3. Before you begin that chapter, take a minute to reflect on the various ways in which mathematical reasoning has surfaced in illustrations and explanations in this chapter.

References

Adler, I., ed. 1972. *Readings in mathematics*. New York, NY: Ginn & Co.

Burton, G.M. 1985. Writing as a way of knowing in a mathematics education class. *Arithmetic Teacher* 33 (4): 40–45.

Davidson, N., ed. 1990. *Cooperative learning in mathematics : A handbook for teachers.* Menlo Park, CA: Addison-Wesley.

Davison, D.M., and Pearce, D.L. 1988. Using writing activities to reinforce mathematics instruction. *Arithmetic Teacher* 35(8): 42–45.

Farrell, M. A. 1969. Mathematics of the culturally disadvantaged young adult. *American Mathematical Monthly* 76: 1053–1056.

Farrell, M.A., and Farmer, W.A. 1988. *Secondary mathematics instruction*. Dedham, MA: Janson Publications.

LeGere, A. 1991. Collaboration and writing in the mathematics classroom. *Mathematics Teacher* 84(3): 166–171.

National Council of Teachers of Mathematics. 1989. *Curriculum and evaluation standards for school mathematics.* Reston, VA: NCTM.

Curriculum Standards for School Mathematics

STANDARD 3: Mathematics as Reasoning

In grades 5–8, reasoning shall permeate the mathematics curriculum so that students can

— *recognize and apply deductive and inductive reasoning*
— *understand and apply reasoning processes, with special attention to spatial reasoning and reasoning with proportions and graphs*
— *make and evaluate mathematical conjectures and arguments*
— *validate their own thinking*
— *appreciate the pervasive use and power of reasoning as a part of mathematics.*

In grades 9–12, the mathematics curriculum should include numerous and varied experiences that reinforce and extend logical reasoning skills so that all students can

— *make and test conjectures*
— *formulate counterexamples*
— *follow logical arguments*
— *judge the validity of arguments*
— *construct simple valid arguments*

and so that, in addition, college-intending students can

— *construct proofs for mathematical assertions, including indirect proofs and proofs by mathematical induction.*

Reprinted with permission of the publisher. See acknowledgements.

Chapter Three

Mathematics as Reasoning

Some readers of the *Standards* may consider this section of the document "old hat," since the outcome statements use words and phrases they've always connected with mathematics classrooms. Are they correct? Has learning to reason long been a part of the curriculum? Nineteenth century mathematics teachers thought so. Teachers of that era (and some today) thought that students learned to reason by learning mathematics—more particularly, by learning geometry. One way they tried to ensure that this happened was to have students copy and memorize classic examples of deductive reasoning, such as geometry proofs. Unfortunately, that tactic often resulted in a long-lasting hatred of geometry and little evidence of improved reasoning.

  How much attention does this standard give to facility in proof-writing? What other facets of reasoning are considered? Isn't reasoning a part of our everyday experience? What's so special about mathematical reasoning? If these questions aren't easy to answer, you'll understand why the relationship between mathematics and reasoning has been the subject of every major report on improving mathematics curriculum and instruction since the report of the Committee of Ten in 1899. With these questions in mind, study the nuances reflected by the outcomes under Standard 3.

The Vision: *What's New?*

If national committees have been making recommendations about emphasizing mathematical reasoning for almost one century, what's new about these 1989 recommendations? Perhaps it's not so much a matter of novelty, but of emphasis. At both groups of grade levels, the students are expected to apply inductive and deductive reasoning and to recognize these two forms of reasoning. Thus, one important area of emphasis is to make sure students have experiences in recognizing patterns and making conjectures as well as experiences in defending a generalization by arguing from accepted assumptions or from logical principles. We call the former *inductive reasoning* and the latter *deductive reasoning*. In this standard, the NCTM authors asserted that students can appreciate the dynamic nature of mathematics only if they are given opportunities to develop both kinds of reasoning.

A second area of emphasis is related to these two reasoning processes. The students are to recognize the **difference** between inductive and deductive reasoning. The *Standards* authors took some time to expand on this outcome. Even at the younger grade levels, they asserted, students need to learn that mathematicians look for patterns, conjecture, and test, but do not consider their work finished when they identify a pattern that seems to work. The authors emphasized that students from grades 5–12 need to learn that, in mathematics, there is a necessary next step—that of validation by arguing from accepted principles.

In the case of spatial and proportional reasoning and reasoning from graphs as well as in the case of inductive and deductive reasoning, the *Standards* authors emphasized the interdependence between development of these understandings and the development of the students' intellectual and verbal skills. This is an extremely important part of this standard. Intellectual and verbal development must be discerned and used to design the classroom materials, the mode of student/teacher interaction, and the type of evaluation. According to the authors, college-bound senior high students (from a practical point of view, students in college-track classes) should experience gradually more sophisticated and more formal proof experiences. There is an implicit assumption that these students (1) may need these mathematical experiences in future mathematics courses and (2) will be intellectually capable of learning more abstract mathematics.

Find the words *validity*, *evaluate*, and *validate* in the Standard 3 outcomes. Notice how the complete statements of those outcomes are related to the gradual development of a number of strategies to test a generalization or to disprove an assertion. Middle school students can be shown how to use a counterexample, how to test a generalization, and how to argue that a statement is never true. They can also be helped to understand that one more working instance of an apparent rule does not "prove" the rule. Students at further stages of intellectual development can be introduced to fallacious reasoning patterns, such as reasoning from a converse, and can be helped to write logical arguments. Once again, the *Standards* authors pointed out the need for the gradual development of these concepts and principles, even at the upper grade levels. They also called for the infusion of reasoning activities throughout mathematics, not just in geometry courses.

In the descriptive material following the grades 9–12 material on this standard, the authors referred to two different kinds of real-world applications. The first, the use of logic forms such as contrapositive, modus ponens, and the like, can help students evaluate the arguments found in newspaper ads, political speeches, and a host of other real-world situations. The second, the area of statistical inference, is an applied area of mathematics that is referred to in later standards for both grades 5–8 and grades 9–12. Statistical inference is highlighted in the Standard 3 material because statistical conclusions are written in a way that seems to imply that they are universally quantified. The authors give an example of a conclusion resulting from a study of a sample of men and women : *Men are taller than women.* What is the probable meaning of this statement? How is it different from the universally quantified statement: $\sin^2 \theta + \cos^2 \theta = 1$. (For further background on this question see page 145 in the *Standards*.)

A Bridge to the Classroom

As noted throughout the *Standards*, the authors envisioned a classroom in which students are actively involved in their own learning and thus are led to appreciate the dynamic nature of mathematics. Nowhere does this vision become more apparent than in the implications of Standard 3. If students are to be given the opportunities to develop reasoning skills and understanding, teachers cannot present concepts and rules in finished form and simply expect students to memorize the rules and ways to apply them.

Inductive and Deductive Reasoning

Consider the following outline of a lesson on a geometry rule. In this lesson on the sum of the angles of a triangle, the teacher includes a hands-on activity in which students fold paper triangular shapes so that all vertices meet at a designated point on a side (Figure 3.1).

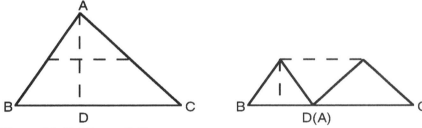

Figure 3.1: Folding activity

Students are asked to fold the shape until vertex A coincides with point D, the foot of the perpendicular from A to side BC. A teacher demonstration with a large triangular shape is used to help clarify the instruction. Then the students are asked to study the resulting shape(s) and make a conjecture to be shared only with partners. Next they are to fold until vertex B coincides with point D. The teacher tells the pairs to study this result, decide whether to change their original conjectures or make a different conjecture. In the final folding process, they move vertex C to point D. In a question-and-answer follow-up, the teacher listens to the conjectures, asks for support for each, and has students focus on the angle/size conjecture. A student may say, "I knew that the other two angles were going to fit next to the first one. The shapes just looked right."

Activities

3-1. How should the teacher prepare the paper shapes so that students do not limit conjectures to a particular kind of triangle?

3-2. Other than the kind of triangle, what other limitations might be inadvertently built into a set of shapes?

3-3. Design questions that would help students focus on angle size, reword a conjecture in terms of sum of angle measures, and recognize that the total class set of examples involved diverse triangular shapes.

3-4. What questions or examples might be used to get students to realize that they not only have not tested all cases, but **cannot** test all cases?

Notice the subtle problem alluded to in Activity 3-4. At the middle school level, students can be made aware that there are limitless variations on the size and shape of triangles. They can be helped to understand that there is no way they can test all cases to **prove** the conjecture. Students who are more abstract reasoners should be asked to reflect on the differences between the physical shapes and the concept of *triangle*. They would be aware of a different reason why this or any other measuring activity can only suggest a conjecture or an approach, but not result in a proof.

What is the middle school teacher to do in order to alert students to the necessity of proof when they do not possess the prerequisite skills to understand an appropriate proof? There is no easy answer. The teacher can indicate that such a rule is proved in later mathematics courses and that proof requires skills not learned in the current course. Most students will accept such a statement, as long as proof is not always characterized as a process beyond their comprehension.

However, some teachers are uncomfortable about unfinished agendas, such as letting students know they have not proved, and at this point do not have the skills to prove, a generalization. These teachers would fail to raise the issue of whether the assertion has been proved and, instead, begin using the assertion as if the hands-on activity were sufficient. By not considering the limitations of inductive reasoning in class, students may learn an incorrect concept of proof.

Activities

3-5. Read Mr. Cain's lesson on the concept of π on pages 153-154 in the Reader.

a. How did he (1) choose materials to give a broad range of examples on which to base a conjecture and (2) treat the question of whether the students had proved the conjecture?

b. What principles relating to the probable intellectual development of his students seem to be reflected in (1) his lesson design and (2) the level of abstraction he used in his lesson?

c. Identify some of the nuances relating to π that would need to be considered by his students at later stages in mathematics courses.

3-6. Refer to the algebra lesson on page 155 of the Reader. Complete the example for the powers of 11. Why is it important to have students experience the fallibility of inductive reasoning?

Providing students with experiences in inductive reasoning is not only important to their understanding of, and appreciation of, the structure of mathematics, it also makes good instructional sense. Formulas, theorems, and algorithms are examples of rule learning, and meaningful learning of rules is intrinsically related to student construction of the chain of related concepts. (For further background material on rule learning, see pages 137–138 in *Secondary Mathematics Instruction*.) The instructional strategy pattern on page 169 in *Secondary Mathematics Instruction* includes the conditions for rule learning. The condition directly related to inductive reasoning experiences is found in the statement

> ...*cue (via questions, lab work, applications) students to find the pattern by chaining concepts; get the rule stated (by students, if possible)....* (Farrell & Farmer, 1988, p. 169)

However, there's another reason such lessons make good instructional sense; it has to do with retention. We know that it is inefficient to reteach and review because students have "forgotten" a rule. Some would say that the rule might have been stored in short-term memory, but not in long-term memory. What components of the angle-sum lesson might lead to long-term retention? Did you consider the potential of that lesson for enhancing cognitive and social interaction? The effects of such lessons on retention lay to rest the argument: *But it takes too much time to have that kind of lesson! I could give them the rule and have them doing examples related to it in half the time.* The arguer is correct about the short-term time advantage. However, when the long term is considered, the argument isn't relevant. When material is "covered" by the teacher in the way described by the arguer, meaningful learning suffers and both retention and transfer are poor. If that teacher were to measure the time used for review of forgotten ideas and for correction of errors, the time advantage would tilt in favor of the active student lesson. Moreover, student motivation tends to be more favorable when students are involved in a more active mode and is more negative when faced with typical review sessions.

Another way of helping students understand the difference between deductive and inductive reasoning is to consider the role of these reasoning types in science. All middle school students take science and many students in grades 9–12 take at least one course in a branch of science. They are aware that there are some commonalities in the content they study. Depending on the students' experience in these courses, statements such as

> *The electric current in a circuit is directly proportional to the electromotive force and inversely proportional to the resistance.*

can be used to help students examine the way in which scientists would verify scientific assertions. Notice that the concepts in this particular assertion, like those in all mathematical assertions, can only be modeled in some way, whereas some scientific concepts (*e.g.,* flower) have concrete exemplars. However, unlike mathematical assertions, this assertion cannot be **proved** now or in later science courses. In science, the test is verification in the Real World. How well do the observed or measured relationships among the data correspond to the principle? If there is disharmony between the empirical results and a theoretical principle, it is the theoretical principle that will be modified or discarded, or the logical reasoning examined for flaws. "Valid and repeatable data can never be disregarded." (Farmer, Farrell, & Lehman, 1991, p. 88)

Since the historical development of mathematics and science is intimately related, readings in the history of both subjects can advance student understanding and help promote **Mathematical Connections**, the subject of Chapter 4. Senior high students can be assigned library projects in the work of scholars such as Copernicus, Ptolemy, and Kepler in order to understand some of the influences that eventually led to the replacement of the geocentric theory by the heliocentric theory. They may be surprised to learn that some scholars, as well as many others, refused to consider the alteration or rejection of the geocentric theory, despite the presence of contradictory data! If students read about the work of Newton and Leibniz, they will learn that these inventors of calculus, as well as many subsequent mathematicians, used new mathematical concepts that were not self-evident, primarily because these concepts led to real-world solutions. The ancient deductive logic, based on universally accepted principles, was set aside for a time. These examples of human struggles and errors in the development of mathematics and science help students realize that the concept of appropriate validation developed over time, as humans made decisions about the nature of their attempts to understand and appreciate the world.

Instructional strategies

"So class, a generalization resulting from inductive reasoning is not proved, unless it can be derived from previously accepted concepts and principles." Do we just reject such generalizations, if we're unable to prove them? Not if our study of history has merit! In the history of mathematics, many generalizations first existed as conjectures based on a number of instances. How can teachers give concrete operational students some sense of the need for, and nature of, proof?

Activities

3-7. Refer to Ms. Pauli's lesson on multiplication of decimals on page 156 of the Reader.

a. Analyze Ms. Pauli's misuse of the word *prove*. Was it wrong for her to have asked the class to test Bob's rule with some other cases? Why or why not? How else could she have summarized the results of the testing?

b. Talk through the suggested argument Bob might have given if Ms. Pauli had asked him to defend his rule. In what ways could this argument be defended as a proof?

3-8. Refer to the honeycomb lessons on pages 158-160 of the Reader.

a. Suppose that the students are manipulating tiles and recording data in a table, such as Table 10.3 from *Secondary Mathematics Instruction*. What reasoned argument could a student provide for reaching the conclusion that all possible tilings that satisfy the given conditions have been found?

b. When students decide on capacity advantage by eyeballing the respective regions of each constructed cell, have they completed a proof? Why or why not?

c. If the method used to decide on capacity advantage consisted of computing the area of each cell for a fixed perimeter of t and comparing these measures, would that method constitute a proof? Why or why not?

How does a teacher decide when to ask students to justify a conclusion? Given the evidence from Ms. Pauli's lesson, it would seem that justification of some sort should be a natural next step. In fact, a student might wonder why one should bother to test at all. Why not just attempt to prove the generalization? How would you defend the value of testing, as well as the need for justification?

The *Standards* authors speak of *reasoned arguments*. Middle grade teachers need to be able to help students realize the nature of a reasoned argument and its stance as an informal proof. In the next example, middle school students use calculators to look for patterns and are then guided toward an informal proof of a conjecture.

Sums and Products of Squares

a. Use your calculator to fill in the following table. Look for patterns.

Add	Sum	Multiply	Product
2^2 and 3^2		2^2 and 3^2	
2^2 and 4^2		2^2 and 4^2	
2^2 and 5^2		2^2 and 5^2	
2^2 and 6^2		2^2 and 6^2	
2^2 and 7^2		2^2 and 7^2	

    b. Continue the table for three further pairs of numbers and guess what the sums and products would be. Check your guesses on the calculator.

When the students have checked the conjectures, the teacher can focus on the pattern found by multiplying two squares—*i.e.*, the product of two squares is a square. Students can be helped to defend this generalization by asking them to rewrite the product in terms of factors, as in the following.

> *What do we mean by a square of a number, for example, 5^2 ? Let's examine our indicated products to see if we can find a clue to the final result being a square. How could we rewrite $2^2 \times 5^2$ in terms of its factors? Examine the computed product of 100 and let's rewrite that as the square of a number. Now does anyone see a way to rearrange the factors in the indicated product [2 x 2 x 5 x 5] so that the final product [10^2] is reflected?*

Since this sequence of questions and responses is in terms of a particular pair of numbers, the students can be asked to use the same process on other pairs. Some students will need to examine multiple pairs before they are ready to defend the idea that the same process would work for any two pairs of squares. Other students will move to this kind of reasoned argument more swiftly. It is important to allow students to build up a reservoir of needed concrete examples before seeking an explanation of the process. This is also an excellent opportunity to have students explain their thinking to one another. **Mathematical Communication** in this instance has students helping each other over developmental stumbling blocks in reasoning. Finally, this kind of proof-constructing activity helps to alert these students to the deductive side of mathematics.

    While middle school teachers are expected to help students develop the concept of informal proof, high school teachers are being asked to broaden their notion of proof beyond that of a formal, two-column listing. "You mean that a proof doesn't have to be written in two columns, headed *Statements* and *Reasons* ?" No, it doesn't. In fact, a paragraph-style proof may make much more sense. Is rigor thrown out the window? No, just re-examined and applied differently. These more liberal attitudes with regard to the nature of proof may take some getting used to.

Activities

3-9. With a partner, review a well-known theorem from geometry, such as *The base angles of an isosceles triangle are congruent.*

Take turns giving each other verbal arguments for the truth of the theorem. Discuss any difficulties or issues related to these verbal arguments.

3-10. Independently, write paragraph-style proofs of the theorem used in Activity 3-9. Exchange papers and analyze each other's attempt. Is there a reasoned argument? Is there an appeal to previously derived principles or concepts? Are there logical transitions so that the set of sentences reads well and makes sense?

Most students find it more difficult to **write** a satisfactory proof, than to **verbalize** one. Was this the experience of you and your partner? Unfortunately, some teachers seem to reject the validity of an oral argument and to place the label *proof* only on a particular kind of written statement. Reflect on the following statements about proof doing and proof writing.

> *Secondary school students who can do a proof but cannot yet write a satisfactory proof need to know that they have good problem-solving skills, even though they must work at the mechanics of proof writing. Prospective teachers need to learn that during proof doing, rigor and precision take a back seat to the generation of ideas.* (Farrell, 1987, p. 245)

Thus, proof doing may occur during a question-and-answer session led by the teacher, by oneself, or in small groups. In all cases, the signal that a proof has been completed is the consensus—of the whole class, the individuals, or the small group—that a way has been found to derive the final conclusion from the given conditions. (For more background on the relationship of doing and writing proofs to attitudes in geometry and to problem solving, see pages 243–247 in Farrell, 1987.)

Mathematical Induction

In *Standards* classrooms, college-bound students are expected to learn how to prove appropriate statements by the method called *mathematical induction*. The process is especially important because it allows deduction to be applied to a generalization asserted to be true for an infinite set of cases. The process is also difficult to understand. Students sometimes think the instructor is just "waving a magic wand" when writing out a proof by mathematical induction. It just doesn't seem to follow the tight step-by-step patterns of synthetic geometry. Of course, the students are correct that mathematical induction doesn't follow these same proof patterns. It depends on a different set of premises. Why is the application of these premises so difficult to understand? How can the teacher help students comprehend the nature of the application?

Let's consider the sequence of steps in the process. First, the generalization is shown to be true for some value of the variable—often for the initial value of the variable. Second, the generalization is assumed to be true for a value, k. Finally, the process uses the statement for the assumption for k to deduce the generalization for the $(k + 1)$th value of the variable. Here's one example of the process.

Prove: that the sum of the squares of the first n positive integers is $(1/6)(n)(n + 1)(2n + 1)$.

Proof:

(1) For $n = 2$, the sum of the squares is $1^2 + 2^2 = 1 + 4 = 5$.

For $n = 2$, $(1/6)(n)(n + 1)(2n + 1)$
$$= (1/6)(2)(2 + 1)(4 + 1)$$
$$= (1/3)(3)(5)$$
$$= (1/3)(15)$$
$$= 5.$$

Thus, the assertion is true for $n = 2$.

(2) Assume the truth of the assertion for $n = k$.

i.e., Assume $1^2 + 2^2 + 3^2 + 4^2 + ... + k^2 = (1/6)(k)(k + 1)(2k + 1)$.

3) Then, for $k + 1$,

$1^2 + 2^2 + 3^2 + 4^2 + ... + k^2 + (k + 1)^2$
$$= (1/6)(k)(k + 1)(2k + 1) + (k + 1)^2$$
$$= (k+1)[k/6(2k + 1) + (k + 1)]$$
$$= (k + 1)[2k^2/6 + k/6 + k + 1]$$
$$= (k + 1)[2k^2/6 + 7k/6 + 6/6]$$
$$= (1/6)(k + 1)[2k^2 + 7k + 6]$$
$$= (1/6)(k + 1)[(k + 2)(2k + 3)]$$
$$= (1/6)(k + 1)([k + 1] + 1)(2[k + 1] + 1).$$

The generalization has been derived for the sum of $(k + 1)$ terms of the sequence.

Therefore, by means of steps (1), (2), and (3), the generalization has been proved to be true for the sum of n terms of the sequence.

As you examine the proof, you can appreciate why high school students react with some skepticism the first time they are exposed to the process. A common response is: "But you used something you did not prove—the statement about the k th sequence of terms. I just don't see how you can call that a proof!" Hidden in the confusion over the process is an incomplete understanding of the meaning of the conditional—if p, then q. Students often confuse the truth of the conditional with the truth of the antecedent or consequent. It's hard for them to understand that when a conditional is true, it means that it's possible to "get from" p to q in some logical way. Recall the truth table for the conditional. There are three circumstances when the conditional is true; but only one of these involves a true antecedent. Thus, in proof processes, we simply consider what happens when p is true. If this line of reasoning sounds like hypothetico-deductive reasoning, you've put your finger on the cognitive level of this kind of content. The ability to produce a deductive proof requires formal operational reasoning. (For information on some research in this area, see page 63 in *Secondary Mathematics Instruction*.)

Yet, even formal operational students may need to have some concrete background in order to understand this highly abstract methodology.

Hint: Draw a portion of a staircase and tell students that it is a staircase with an infinite number of steps. Then give the following explanation.

When n = 1, the sum of the squares is just the first square, 1^2, and we can check to see if 1 is equal to (1) (1 + 1)(2[1] + 1)(1/6). It is. Thus the generalization is true for n = 1 and we are on the first step of the staircase (Figure 3.2).

Teacher places a mark on step 1.

Can we get on the second step, the third step, and so on? We could, of course, just keep on testing for n = 2, 3, ..., but eventually we'd have to admit that we couldn't test all cases. Instead we ask, can we show that we can always move from a lower step to the next higher step? If we can show that, and if we can show that we are able to get on a specific step, then we've proved that we can move from that step to the next. That move puts us on another step and the process lets us move up again, and so on.

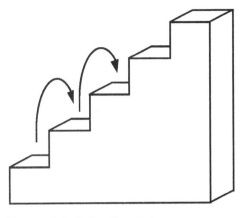

Figure 3.2: Induction staircase

Marks steps with curved arrows.

This is what we've accomplished with our proof on the sum of the squares. We tested for n = 1 and n = 2. (Sometimes it is wise to test for cases that are "deformed," such as a sum with only one addend.) These tests put us on the staircase at step 2. Then we assumed we could get on some step, k, and showed that we could move from any step, k, to the (k + 1) step. That derivation completes our proof. The "move" will take us from the second step to the third; but once we reach the third step, the "move" will take us to the fourth step, and so on.

Notice how the explanation given by the teacher uses a concrete image and analogy to help students with the abstract character of the proof by mathematical induction.

There's another reason why mathematical induction seems like magic. Sometimes text writers provide students with lists of generalizations to prove by mathematical induction, but never involve students in the activity of formulating generalizations. The omission leaves students in the position of assuming that the generalizations were somehow written down in a flash of insight. Without some experience in the inductive reasoning that led to these generalizations and an understanding of the resulting tentative nature of the generalizations, students are reinforced in the belief that proof doing is one of the gimmicks mathematicians made up to terrorize students, rather than a necessary part of the system. Students need to become privy to struggles of mathematicians as they worked to find patterns and to translate them into a useful model.

An earlier activity found in this chapter involved a calculator example in which pairs of squares were added and multiplied. However, the pattern accessible to middle school youngsters was simply that the sum of two squares was not a square, while the product of two squares was a square. They weren't expected to formulate a symbolic relationship that could be used to predict. In contrast, students capable of understanding mathematical induction are also capable of generating some of the conjectures they will later prove. However, they need some cues on how to proceed. It's helpful to begin with a fairly easy example so that an organizational procedure might be shown. Here is such an example.

Find a generalization, if one exists, for expressing the sum of n sequential odd numbers.

Students can be shown how to organize their data in columns—a column with the various values of *n*, a column showing the indicated sum of the positive odd integers, and a column with the computed sum.

Table 3.1   First Stage for Sum of *n* Odd Integers

n	Indicated Sum	Sum
1	1	1
2	1 + 3	4
3	1 + 3 + 5	9
4	1 + 3 + 5 + 7	16
⋮	⋮	⋮
k	1 + 3 + 5 + 7 + ... + (2k − 1)	–

Now the students should be encouraged to add a fourth column, with the sum expressed in terms of *n*.

Table 3.2   Second Stage for Sum of *n* Odd Integers

n	Indicated Sum	Sum	Sum in Terms of n
1	1	1	1^2
2	1 + 3	4	2^2
3	1 + 3 + 5	9	3^2
4	1 + 3 + 5 + 7	16	4^2
⋮	⋮	⋮	⋮
k	1 + 3 + 5 + 7 + ... + (2k − 1)	—	–

Students need to be warned that this last column is not always as easy to complete. There is often considerable effort expended to rewrite the results in terms of the corresponding *n* in a way that reflects a vertical pattern.

Hint: One tactic that can be used to identify the degree of the desired algebraic expression is the delta tactic. Find the differences (deltas) between last column results and the differences between the differences until a common difference occurs. In this case:

$$1 \quad\quad 4 \quad\quad 9 \quad\quad 16 \quad\quad 25 \quad\quad 36$$
$$3 \quad\quad 5 \quad\quad 7 \quad\quad 9 \quad\quad 11 \quad : \Delta$$
$$2 \quad\quad 2 \quad\quad 2 \quad\quad 2 \quad : \Delta\Delta = \Delta^2$$

The common difference is reached in two steps. Thus, the desired algebraic expression, in *k*, is a quadratic.

Activities

3-11. Use mathematical induction to prove the assertion that the sum of *n* consecutive positive odd integers is the square of *n*. Talk with a colleague about the problems, if any, that secondary school students might have with this proof.

3-12. Work out a table as part of an exploration to find an expression for the sum of the squares of the first *n* positive integers. Use the delta tactic to see if there is a common difference at some level. With a colleague, work toward the generalization given in the proof shown earlier in this section.

The difficulty level of Activity 3-12 contrasts sharply with that of the inductive reasoning needed to find the generalization for the sum of *n* positive odd integers. The processes you and your colleague used in connection with this activity are certainly good ways to see **Mathematics as Problem Solving**; thus, this is another way to teach toward more than one standard. It's important to have your students experience these processes and get some sense of the mathematician's point of view when trying to find a pattern. What are some tactics that might have been tried? Have your students respond to the following questions.

> If a mathematician is unable to discern a pattern, does this mean there is no relationship? If a mathematician discerns a pattern that works for many terms, does this mean it will work for all terms?

> Hint: As in the case of all problem-solving work, the discernment of a possible pattern might not be found by some algorithmic procedure. Thus, it is important for the teacher to convey the expectation that there will be considerable trial and error, possible dead ends for some students on some examples, and the need for persistence and restructuring. For example, if the delta method and its application do not seem to lead to a solution, students might try rewriting the indicated sums in factored form and examining these expressions for the appearance of the related *n*. This is one example of restructuring the problem material. Another tactic that sometimes helps is to work with the indicated sums and search for another way of rewriting them. A third tactic is to look for a relationship between adjacent sums. This tactic is sometimes referred to as: *What's next?* The finding of a vertical relationship between adjacent sums sometimes helps clarify the nature of the horizontal relationship between a sum and the number of terms being summed.

There is ample evidence that the understanding of, and ability to use, mathematical induction and these complex instances of inductive reasoning require a high level of cognitive development. Similarly, there are developmental considerations associated with the appropriate use of both proportional and spatial reasoning. Each of these will be considered next.

Proportional Reasoning

Although the NCTM authors specifically mentioned proportional reasoning only in the grades 5–8 section, it is important to recognize that there is a need for further attention to the development of this kind of reasoning throughout the 9–12 grades. In this case, we have a set of competing pressures. Some aspects of the topic of ratio and proportions are found in mathematics and science courses throughout the middle and high school years. At the same time, researchers have found that there are conceptual gaps and deficiencies in proportional reasoning abilities, even among senior high students in college-track courses. (For background on this area, see pages 61–62 in *Secondary Mathematics Instruction*.) It appears that proportional reasoning requires repeated experiences of a concrete nature and that the teacher must be alert to the appearance of misconceptions.

Activities

3-13. Refer to the bicycle gear lesson on pages 147-148 in the Reader.

a. What features of this lesson would help concrete operational students develop understandings about ratio and proportion?

b. What use is made of inductive reasoning in this lesson?

c. What are the specific advantages of using a large bicycle wheel as a demo and then of having students construct their own cardboard models of gear systems?

3-14. Refer to the lesson based on *The Hobbit* on pages 161-162 of the Reader.

a. In this lesson students were asked to compute heights and ratios of dwarfs. What principles from cognitive development might explain why students would consider these kinds of questions as meaningful?

b. Identify the different levels of complexity found in the set of questions.

c. How might the accompanying map serve as another concrete aid for the students?

d. Use any standard middle school text with a unit on ratio and proportion and locate a set of "word problems" on the topic. How do these compare with the questions in *The Hobbit* lesson? Be especially alert for the use of familiar context, and concrete or visual images.

3-15. You can use the overhead projector and a stencil of a triangle to develop the concept of similar triangles. The image projected on the chalkboard can be matched to a cardboard shape congruent to the stencil. The students will be able to observe that the angles of the smaller cardboard shape correspond exactly with the angles of the larger projected image.

a. Cut out a stencil of a scalene triangle and a congruent cardboard triangular shape. Try the demonstration and write down the kind of question you would ask, or direction you would give to, students. What should they observe?

b. How would you develop the two features of similar triangles—congruence of angles and proportionality of side lengths?

(Suggestion: Move the projector away from, and toward, the chalkboard and have a student trace each image with different colored chalk.)

c. What features of the demonstration are likely to promote cognitive interaction?

The mathematical language in the questions from the *Hobbit* lesson makes it clear that the teacher is building on a previous introduction to ratio, while the bicycle gear lesson is alluded to as an introduction to the concept of ratio. Indeed, it is suggested that questions about the relationship of gear teeth to revolutions might not even use the label *ratio* at first. These lessons reflect a shift from hands-on activity to paper-and-pencil work with an imaginative story line. Even in the latter case, visualization is deemed important and is provided in the form of a map so that the travels in the story can be followed in a concrete way.

Read about the Mr. Short/Mr. Tall puzzle in the Reader on page 163 before continuing. Because a ratio is often written in fraction form and a proportion is written as a pair of equivalent fractions, students often fail to realize that the concept of ratio is not the same as that of a fraction. The significance of the concept for comparing two entities that are related in a multiplicative way is not realized. In the Mr. Short/Mr. Tall puzzle, the reader is told that the height of the pictured Mr. Short is 4 buttons and that of Mr. Tall is 6 buttons. The buttons are not pictured. Then, the reader is asked to measure the height of Mr. Short with a paper clip chain and to conjecture the height of Mr. Tall in paper clips. A surprising number of students who have been taught ratio in several courses will use an additive strategy. That is, since Mr. Tall is 2 buttons taller than Mr. Short, they would say that Mr. Tall should be 2 paper clips taller than Mr. Short. Farrell and Farmer (1985) found this same strategy used by secondary school students in college-bound mathematics and science courses when faced with an analogous novel situation. Many of these students disregarded the possibility that the changes might be multiplicative rather than additive. Thus, the first conceptual issue faced by teachers is to provide examples of real-world changes, such as the bicycle problem, in which the relationship is clearly multiplicative.

One way of contrasting multiplicative and additive situations is to engage middle school students in a lesson using *linkages* (if your classroom has the proper technology, this activity can be done with graphics or geometry software). Linkages are rigid strips that are connected with fasteners that create joints. A carpenter's ruler is a series of such linkages. In this lesson, our strips will be the holed edges of computer paper—not very rigid, but they'll do. Paper fasteners will be used to link strips together through existing holes. Student groups should be given a collection of paper fasteners and strips and told to consider the distance between two adjacent holes as a unit distance.

First they should be asked to construct a triangle with sides of 3 units, 4 units, and 5 units by linking three strips appropriately. Long strips can be cut so the figure will be less cumbersome. A demo triangle placed on the overhead projector table will help to clear up any problems that groups might have. Next have the students calculate side lengths obtained by adding 2 (then 3 and finally 4) to each of the side lengths of the original triangle. Have them work together to calculate the side lengths obtained by multiplying 2, 3, and finally 4 by each of the sides of the original triangle. When the family of these triangles has been constructed, students should be asked to observe what changed and what remained the same. Have them record the side lengths of each triangle in a chart, find and record the perimeters of each triangle, and conjecture results for larger triangles formed by multiplying the lengths of the sides of the original triangle by a different constant. Conjectures can be tested by using the links to construct the figure in question. Finally, ask students to draw the family of triangles on dot paper. The triangles might be labeled as the *x2*, *x3*, and *x4* triangles, with the 3, 4, 5 triangle labeled as the original triangle.

Next, have the students calculate side lengths obtained by adding 2, 3, and 4 to each of the side lengths of the original triangle. Have the groups construct these triangles with linkages, and compare these shapes with the original triangle. The students should again be asked to observe what changed and what stayed the same, record side lengths in a chart, find and record perimeter for each triangle, and conjecture results for larger triangles formed by the same process. This family of triangles should be drawn with a different color pencil than those of the multiplicative family, but on the same dot paper as before. The triangles should be labeled as the *+2*, *+3*, and *+4* triangles. The teacher can help consolidate student thinking, draw out observations to be shared, and emphasize important conclusions.

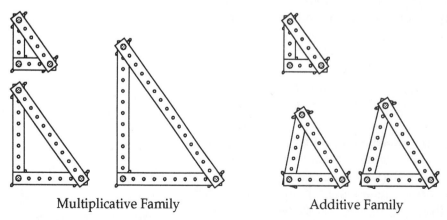

Multiplicative Family   Additive Family

Figure 3.3: Linkage triangle families

Activities

3-16. Respond to these questions on the basis of the Linkage lesson.

a. The original figure was a 3, 4, 5 triangle. What advantage is gained from using this as a starter triangle, rather than a scalene triangle with three acute angles? Consider the advantages or disadvantages of using a rectangle with length 5 and height 3 as an original figure.

b. In the case of each family, the student is asked to generalize about triangles not yet drawn. As an extension for mature students, ask them to generalize for an addend of n and for a multiplier of n. Complete the table in each case and discuss with a partner the possible reasoned argument that a middle school youngster might give for each generalization.

c. Some students may conjecture that the additive family is beginning to look like a collection of equiangular triangles. How can you help middle school students to deal with the "evidence" of their eyes? In what way could secondary students use their knowledge of trigonometry to explain what is happening?

3-17. Consider Activity 3-15, in which the concept of similar triangles is introduced. What is the relationship of the changing distance between the projector and the chalkboard to the changing size of the image triangle?

3-18. Positive Sally says that a snapshot and its enlargement are related in an additive way. Design a test that students could use to decide on the truth of her assertion.

The linkage lesson has the potential to serve as a problem-solving activity, as well as an activity emphasizing proportional reasoning. Be sure to construct a set of these linkages before you try this lesson. Then it will become clear why it is important to have the students complete the multiplicative family first. With their attention firmly fixed on the obvious similarity of shape of the new triangles to the original triangle, they will be more likely to look for the same kind of relationship when they begin to construct the additive family. If you reversed the process of construction, conjectures about the possibility of equiangular triangles might draw attention away from the proportionality theme. After the students have talked about the two effects of the multiplicative process for triangles, conjectures about the additive family can be addressed. Secondary students could use calculators to explore side ratios, not only for the constructed triangles, but for triangles formed by using very large addends. They could also be asked to consider the case where the original triangle is an equilateral triangle, an isosceles triangle, or a scalene triangle with sides that are not consecutive integers. Since proportional reasoning is so intimately related to cognitive development, extensions of this kind are important classroom activities throughout the secondary years.

Teachers need to be alert to real-world examples of proportional objects, such as nests of boxes, most coins (a few are polygonal in shape), model planes and cars versus the real versions, TV images on 5", 13", 20", and larger screens, and so on. They also should help students realize that just because a real-world object increases in a multiplicative way in one dimension, it doesn't necessarily increase in all dimensions in this same way. Give students the following problem.

	length (height)	*weight*	*waist*
Tireless Tom when 3 mo. old	25"	15 lbs.	19"
Tireless Tom as a 7th grader	5'4"	?	?

The table shows Tireless Tom's dimensions at age 3 mo. His dimensions were about average for a 3 to 4 month old baby. By seventh grade, Tireless Tom had grown to a height of 5'4". (How tall are you? Your partner? Other seventh graders?) Tireless Tom's height in seventh grade is how many times his height when measured in infancy? (Round to nearest whole number.) If his weight and waist size have also multiplied by the same amount, what would each of these measures be? Fill in the table with these measures. Talk with a partner about the data. What would Tireless Tom look like? Compare Tireless Tom's dimensions with your own and with those of your partner. Talk about this problem and what the results suggest about growth changes.

After considering this problem, students can be given data on real-life giants (consult the latest *Guinness Book of World Records*) and led to consider the probable reasons for their limited life span. If they have read *Gulliver's Travels*, they can be asked to consider the conclusions of the Lilliputians who assumed that Gulliver would need 12^3 as much food as they because he was 12 times taller than they were. Students can be helped to realize that the Lilliputians were applying proportional reasoning, but were doing so for an inappropriate situation. As these few illustrations show, opportunities for teaching **Mathematical Connections** to literature and to science abound when studying ratio and proportion.

Each of the activities and examples given thus far portray ratio in a geometric context and make use of spatial reasoning. However, ratios appear in a variety of other contexts. If we drive at an average speed of 62 mph, how far will we have gone in 4 hours? The octane in gasoline, the percentage of concentrate in fruit juices, the purchase of 3 gumballs for 25 cents, the maximum rate at which a motorbike travels are just a few of the other contexts in which ratio and proportion are applied. The transition from ratios in which a unit is compared to some constant (*e.g.*, 1: 2, 1: 3, 1: 4) to those in which the comparison is based on a *hidden* unit (*e.g.*, 2: 3, 3: 5, 4: 7) is a complex developmental problem. Take another look at the Mr. Short/Mr. Tall puzzle. Here we have an example of a comparison of

Mr. Short's height and Mr. Tall's height to each other, based on the relationship of each height to the hidden unit. The ratio of their heights is 2:3. That is, Mr. Short's height is twice the unit and Mr. Tall's height is 3 times the unit. The strategy for approaching such problems is not simply a matter of reducing fractions, but noting the presence of the hidden unit. Can you see why these types of ratios have been so difficult to understand? What principles from cognitive developmental psychology help to explain the complexity of the problem? Why has the strategy of simply treating these ratios as fractions not been a successful one?

Activities

3-19. The ratios in the puzzle problem, the triangle transformation activity, and the baby measures activity are used to compare the measures of like objects (*e.g.*, two heights, two perimeters, two waist measures). How can rates, such as mph, be explained as a ratio?

3-20. Develop a data table for possible heights of Mr. Short and Mr. Tall, using the 2:3 ratio. What is the hidden unit? How might such data tables be used to help students understand the meaning of a proportion and how to find a missing term?

Spatial Reasoning

When observations of shape characteristics are being sought, as in the Linkage activity, students are being asked to use an aspect of spatial reasoning. Spatial abilities are diverse and there is some evidence that at least some of these abilities can be improved by instruction. You may have read some of the news stories about research on the relative ability of boys and girls in spatial reasoning. There is a growing body of work in this area and some evidence that boys may have a head start over girls in spatial abilities as a result of differences in early play materials. (See page 65 in *Secondary Mathematics Instruction* for an introduction to this area of research.) In what sense is spatial reasoning a part of cognitive development? For example, what makes the understanding of volume of a three dimensional figure dependent on formal operations? What, in fact, does a volume number, such as 33 cm^3 mean? (If you need help on any of these questions, see Chapter 3 in *Secondary Mathematics Instruction*, specifically pages 52–54.)

   Here are just a few of the kinds of spatially-biased topics that need special attention from the teacher: the application of formulas for area and volume, the identification of relationships given overlapping or contiguous shapes, the relationship between 2-dimensional pictures and their 3-dimensional reality, the effect of dissection on a shape, the effect of transformations—such as translation, rotation, reflection, and dilation, and graphing and concepts related to graphs. In all of these areas, the overriding principle is to immerse the students in concrete activities. Why? What principles of meaningful learning and cognitive development support such an assertion?

Activities

3-21. Refer to the feltboard or velcro board idea on page 164 of the Reader.

a. Assume that your students have learned the derivation of the area of a parallelogram and seen the formula developed by use of shapes on the velcro board. You decide to pose the problem of finding a formula for the area of a trapezoid by similar dissection and rearrangement. What questions would you ask the students? How might they be expected to show the process they used?

b. Here are the ways two different students solved the problem posed in **3-21a.**

Student #1 : *I cut a trapezoid into a rectangle and two triangles so the area is the sum of the three areas.*

Student #2 : *I did the same thing, but then I moved the two triangles together to form one larger triangle, so the area is the sum of the area of the rectangle and the area of the new triangle.*

How could you work with the responses of these students and the velcro board to move all toward a formula?

3-22. Refer to the sunflower seed lab activity on page 165 of the Reader.

a. Explain how the growing, 3-dimensional bar graph provides a conceptual basis for an understanding of the normal curve.

b. Why is this sunflower seed array an example of a bar graph, rather than an example of a histogram? What aspect of the activity needs to be emphasized to clarify this distinction?

3-23. In this activity, you are given an opportunity to try out your own spatial abilities. Begin by studying the shell pattern of the chambered Nautilus.

a. Then construct a rectangle with sides 8 and 13 units. Using one of the short sides of the rectangle, form an 8 by 8 square within the rectangle. The original figure is now divided into a square and a new rectangle, 5 by 8. Within the smaller rectangle, form a 5 by 5 square, thus yielding a new rectangle, 3 by 5. Continue to form squares within the new rectangles. Study the sequence of rectangles, in terms of the ratio of their sides.

b. Label one vertex of the original rectangle. Locate and label the vertex that corresponds to that in the next rectangle, and the next, and so. Notice how a given rectangle appears to shrink and rotate to form the next rectangle? Trace the spiral formed by connecting corresponding vertices of the sequence of rectangles. Compare your result with that in Figure 3.4.

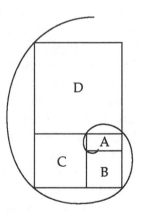

Figure 3.4: Spiral

c. For a true match to the spiral modeled by the Chambered Nautilus, the rectangles would have to be similar, and the ratio of the sides would have to be the golden ratio, $(1 + \sqrt{5})/2$. How are the lengths of the sides of the drawn rectangles related to the golden ratio?

In the last set of activities, secondary students (and you) are engaged in various kinds of concrete experiences—cutting and moving a set of figures to form a new shape, selecting particular seeds by observing some distinguishing characteristic, and constructing shapes on paper. In the first activity, the transformation into a known shape or set of shapes is just the first step toward generating a new formula. The teacher may need to help the students denote lengths by symbols and reduce the number of symbols to make the formula as simple as possible. In the next two activities, the lab or construction work is ongoing and early observations may lead to erroneous, but interesting, conjectures. The constructions in both cases evolve as more steps occur and, thus, can lead to further observations. In order that development of spatial abilities may be encouraged, it is important that the teacher's early instructions are open-ended (*e.g.*, Find a pattern) so that many aspects of the results of the activity will be considered. The teacher's role is to allow the students to learn from subsequent steps, not to hurry them along. They may need to be cued to expand their observations to consider, for example, the relationships of sides to sides, angles to angles, individual glass containers to each other and to the whole.

Notice how these illustrations differed in the ways in which they engaged the students' spatial abilities. Why is Activity 3-23 appropriate for students who may be transitional or formal reasoners? Why not simply have the teacher show the students static pictures of the evolving rectangles and ask the students to observe the common pattern? What might be gained by student construction of the set of rectangles? If you identified the evolving aspect of the spiral formed from points on the curves, you've also hit on an aspect of the growth pattern of real-world representatives of the equiangular spiral. The chambered Nautilus does live and grow, and its shell must also change in order to accommodate the creature. Notice where the growth begins in the shell and the size and shape of new chambers of the shell. Once again, we have a beautiful vehicle for teaching **Mathematical Connections**. For more on this particular subject, see Chapter 6 in Thompson (1917/1961).

This final example is one more indication of the innumerable possibilities for instruction that simultaneously help students attain multiple standards. The first four standards are not only strands permeating each of the specific content areas; they also interweave with each other. However, that kind of integration, though desirable, is not automatic. It is possible to use the spiral activity without treating the properties of the chambered Nautilus in the real world. It is also possible to have students complete the dissection of the trapezoid and obtain a formula without ever delving into their understanding that area is conserved when they cut and rearrange the shapes. In the latter case, a gap in spatial reasoning may be overlooked. Each of these examples is indicative of the need to be quite clear on the multi-purposes associated with an activity or a strategy design. As we will see when we work on Chapter 5, an understanding of evaluation strategies will help to keep our focus on the desired standards.

References

Farrell, M.A., and Farmer, W.A. 1985. Adolescents' performance on a sequence of proportional reasoning tasks. *Journal of Research in Science Teaching* 22 (6): 503–518.

Farrell, M.A., and Farmer, W.A. 1988. *Secondary mathematics instruction*. Dedham, MA: Janson Publications

Farmer, W.A.; Farrell, M. A.; and Lehman, J. R. 1991. *Secondary science instruction: An integrated approach*. Dedham, MA: Janson Publications.

Farrell, M.A. 1987. Geometry for secondary school teachers. In *Learning and teaching geometry, K–12* , ed. M.M. Lindquist, 236-250. Reston, VA: National Council of Teachers of Mathematics.

Thompson, D'Arcy. 1961. *On growth and form*. Abridged Edition. Cambridge: The University Press.

Curriculum Standards for School Mathematics

STANDARD 4: Mathematical Connections

In grades 5–8, the mathematics curriculum should include the investigation of mathematical connections so that students can

— *see mathematics as an integrated whole*
— *explore problems and describe results using graphical, numerical, physical, algebraic, and verbal mathematical models or representations*
— *use a mathematical idea to further their understanding of other mathematical ideas*
— *apply mathematical thinking and modeling to solve problems that arise in other disciplines, such as art, music, psychology, science, and business*
— *value the role of mathematics in our culture and society.*

In grades 9 –2, the mathematics curriculum should include investigation of the connections and interplay among various mathematical topics and their applications so that all students can

— *recognize equivalent representations of the same concept*
— *relate procedures in one representation to procedures in an equivalent representation*
— *use and value the connections among mathematical topics*
— *use and value the connections between mathematics and other disciplines.*

Reprinted with permission of the publisher. See acknowledgements.

Chapter Four

Mathematical Connections

Chapter titles such as "Structure of Mathematics", "Perspectives of Mathematics Education", or "Mathematics, Science, and Everyday World Interface" seem to be likely places to read about **Mathematical Connections**. However, ideas for teaching students about mathematical connections may also be found in other places, such as lists of sample discussion topics. For example, the discussion topic "Traffic pattern in school corridor" (Farrell and Farmer, 1988, page 12) is based on the students' everyday experiences, but requires that they design a plan to collect data and develop a set of recommendations. To the students, this exercise may not seem much like mathematics. It isn't much like **traditional** mathematics, but it is a good example of the use of mathematical models—the use of simple arithmetic and geometry to reflect a real-world situation. This view of mathematical connections is addressed by the authors of the *Standards*, but they also give another meaning to the concept of mathematical connections. As you read the outcomes under Standard 4, formulate a description of this additional aspect of mathematical connections.

The Vision: *Away with Bits and Pieces!*

When you recall the work of early psychologists on mathematics learning, you should remember that one of the extreme reactions to that work was to subdivide arithmetic into hundreds of facts to be taught and learned. In more recent decades, adherence to these extreme positions has lessened. Yet mathematics continues to be presented in compartments in some classrooms. *Last year I studied algebra; now I'm studying geometry; I don't see any connection.* Not only are connections not made apparent between courses, but connections may also be ignored for related topics within a course. For example, students might be taught the unit on factoring as if it were a totally new topic with little or no reference to its connection to multiplication. In addition to these compartmentalized approaches, mathematics is too often presented only in its final symbolic or idealized form. It's no wonder that many adults think of mathematics as an ivory tower exercise, a static but difficult discipline with little relevance for most of the students.

This separation of the subject into bits and pieces without connection to the real world has deterred learning, misrepresented the structure of the subject, and led to pervasive negative attitudes. The authors of the *Standards*, throughout both the grades 5–8 and grades 9–12 sections of Standard 4, made it clear that mathematics needs to be taught so that students "see it as an integrated whole." They provided illustrations of several approaches that connect one aspect of mathematics to another. One approach involves the identification and use of a unifying concept or principle. In the grades 5–8 section, the authors showed how Pascal's Triangle could be used as a unifying idea to explore relationships in number theory and probability. A second approach requires the identification and use of the structural links between mathematical principles or concepts. The teacher would analyze the mathematical prerequisites for each topic and then develop lessons so that students could use prior learning as building blocks for new learning. In this way a new rule may be shown to be an adaptation of something previously learned. These kinds of connections may sound similar to the learning hierarchies researched by Gagné (see pages 142–143 of *Secondary Mathematics Instruction*).

Some of the outcomes listed in Standard 4 might be thought of as ways to develop these mathematical connections. For instance, when results are described in more than one mode—physical model or algebraic, numerical, graphical, or verbal description—students learn that the various tools for talking about mathematics help to clarify an idea and, sometimes, to suggest new facets of the idea. Recall the linkage activity with triangle families in Chapter 3 of this text. Side lengths were listed in a table for the successive triangles—a numerical representation. The collection of linkages, which showed a set of triangles in a family, provided a physical model. Even the transfer of the linkage results to a static geometric picture on dot paper provided another representation. Each representation alone gives some information on the nature of the relationship being investigated. Together, they offer a richer view of the relationship and may lead the students to suggest new relationships.

Take a few minutes now and identify the outcomes in Standard 4 that refer to the connections of mathematics with other subjects or the everyday world. Did you select the last two outcomes in the grades 5–8 list and the last outcome in the grades 9–12 list? It is easy to misinterpret the intentions of the NCTM authors if one reads only the outcome statements in Standard 4. You could hardly blame a veteran teacher for being dismayed at the apparent expectation that middle school youngsters should be able to successfully handle applied mathematics. After all, one has to know, and be skilled at using, some rather sophisticated mathematics in order to be able to apply it to problems in chemistry, psychology, and the social sciences. However, that level of application is **not** what was intended by these particular outcomes. The NCTM authors are referring to school subjects being taken by the students—subjects such as physical science, physical education, and social studies— and to the everyday world of the students, both in and out of school.

Thus, the study of simple machines in physical science has connections to the study of several functional relationships. Similarly, student involvement in the band or orchestra can be used as a basis for an exploration of connections between music and mathematics—for example, the concepts of frequency, period, and amplitude. (An introduction to some of these connections with music may be found in *Secondary Mathematics Instruction* on pages 275–276.)

Activities

4-1. Review any two standards from standards 5–13 in either the grades 5–8 or grades 9–12 sections of the *Standards*. Look for outcome statements that seem to emphasize mathematical connections to the real world or to school subjects. Review the explanatory material in the chosen sections for illustrations emphasizing mathematical connections.

4-2. For the two standards selected in your response to Activity 4-1, look for outcome statements that seem to allude to connections within mathematics. Again read the explanatory material for illustrations or author explanations that might substantiate your choices.

You and your colleagues should have found multiple instances of outcome statements reflecting mathematical connections. That result shouldn't be surprising. The standard of **Mathematical Connections** is not only one of the four unifying strands in this document; it is, by its nature, intrinsic to all other standards. The NCTM authors repeatedly refer to the need to integrate mathematics, to use various representations to reflect the same set of relationships, and to look for links between and among mathematical concepts. Indeed, one of the important assumptions of the authors is that this kind of integration is more efficient in terms of instructional time than a non-integrated approach. Thus, students can learn more content when mathematical connections are emphasized. In a succinct summary of the importance of teaching for mathematical connections, the authors of the *Standards* list better understanding and appreciation of the power and beauty of mathematics and increased retention and transfer as effects (NCTM, 1989a, p. 149). These are sufficient reasons to explore ways to make this standard a reality in every classroom.

A Bridge to the Classroom

As you have surely realized from your study so far, implementing the *Standards* requires thoughtful design of instructional strategies with attention to the specific outcomes of the lesson. The process involved in implementing Standard 4 is no different in this respect. However, it may be unique in that the implementation of Standard 4 depends on considerable teacher preparation prior to the design of strategies for a particular lesson. What kind of preparation is needed? How does the preparation mesh with that needed for the other standards? What are realistic ways for a busy teacher to plan to work toward the outcomes of Standard 4? These are the questions to be addressed in the following sections of this chapter.

Connections Within Mathematics

Suppose you are preparing to teach a lesson on the quadratic function or a unit on measurement, how do you work mathematical connections into the lesson? First and foremost, you have to identify those connections. This is an obvious assertion, but not very helpful. How do you identify them? You could sketch out a learning hierarchy so that prerequisite mathematical learnings are clear to you. A hierarchy sometimes shows that there are alternate sequences leading to the new rule being taught. Some teachers have drawn rough concept maps of all the concepts in the unit. The map helps them decide which of the connections pictured in the map should be emphasized in class. No matter which of these tools you choose, your first step is to study the mathematics to be taught and analyze its relationship to past and future mathematical learnings of the students.

Activities

4-3. Study the portion of the teacher's concept map found in the Reader on page 166.

a. The word *graphs* has two connectors. What are the differences between these two relationships?

b. The ideas *point slope form* and *slope intercept form* are connected to *linear equations*. How might they be connected to each other in a lesson?

c. Would you connect any of the concepts in a different way? Which and why?

4-4. Study the learning hierarchy found in the Reader on page 167.

a. Why aren't objectives 2a and 2b sufficient prerequisites for learning objective 1?

b. Does it matter whether a student learns objective 2b before, or after, the learning of objective 2a? Talk over your rationale for your answer with a colleague.

c. Check a contemporary algebra text for the sequence in which the solution of first-degree equations and the factoring of polynomials (objective 3a) is presented. Could the sequence be altered without any problem? Should it be?

You probably found that you agree with the concept map referred to in Activity **4-3** in some respects, and disagree in others. As long as neither you nor the constructor of the original map has made any mathematical mistakes, some differences would be expected. The map serves as a rough guide to planning. It's like a suggestion box that can provide the teacher with a potential lesson sequence, areas of emphasis, and directions for the future. A thoughtful concept map does take time; but once completed, it serves as a starter map for future lessons. The author of the map on Coordinate Geometry may have revised it after instruction and probably revised it again the following year, based on other connections not identified at the time.

The same comments are true with respect to the learning hierarchy referred to in Activity **4-4**. Although there are some sequences that cannot be violated without erring mathematically, there are others that can offer a choice from the typical textbook route. These should make good mathematical sense and may make better psychological sense than the typical textbook route. Remember, the logical way to proceed is not necessarily the best psychological way to proceed when it comes to learning.

The construction of useful learning hierarchies takes considerable time. How can a busy teacher possibly do that kind of planning, as well as keep up with all the other attendant paper work? A realistic approach is to identify two or three areas of a course that seem most unfamiliar or that you find difficult to teach and complete a learning hierarchy for each of these areas this year. The following year, these can be revised as needed and other course areas analyzed.

Activities

4-5. Study the section entitled *Course Planning Schema* on pages 168-170 in the Reader.

a. Review the five major course objectives developed by Ms. Aronowitz. Using any contemporary geometry text as a basis, reflect on the scope of her five objectives. Which areas of geometry are included; which, not considered? In what sense might her objectives be considered as unifying themes?

b. Now consider the eight units found in the figure on page 169. Again, match these against the topics in the geometry text you used for your work with **4-5a**. Choose two of the connections symbolized by the curved arrows. How could these connections be used to help students view the involved mathematics in an integrated way?

Did you notice that Ms. Aronowitz's fifth course objective corresponds to outcomes under Standard 4, while her third course objective sounds like an outcome under Standard 3, **Mathematics as Reasoning**? Both kinds of objectives are valuable in helping students learn and retain mathematics. Think of the imagery found in the study of the stance of athletes (*e.g.*, a golfer ready to address the ball or a batter waiting for the pitch) as it is related to the rigidity of the triangle. Students have a better understanding of the concept of structural rigidity after collecting and analyzing such pictures. They also have a real-world connection that seems to be retained and recalled with ease. Similarly, when students analyze the comparative values of a geometric argument, they learn a great deal about deductive logic. These mathematical connections at the analysis level should also help students to retain and retrieve proof strategies. What classroom strategies might be used when working toward Ms. Aronowitz's third course objective? Small group work seems an obvious first choice, doesn't it? In fact, this course objective might well have been what Mr. Potter had in mind in his problem-solving lesson (See page 131 in the Reader).

In one of the preceding activities, you were asked to consider the unifying theme aspect of Ms. Aronowitz's course objectives. Did you refer to the course plan and the way in which each course objective was overlaid on several units? Perhaps you selected a particular course objective and gave examples of the way it could be the goal across several areas of content. Another approach might have been to show how a course objective could have been treated in more depth as the students progressed through the units. All of these are possible ways in which a course objective can serve as a unifier of mathematical content. In contrast, in an early section of this chapter, you were given an example of another kind of unifier—the use of Pascal's Triangle as a unifying theme. That kind of unifying theme— a concept, principle, pattern from mathematics—is a powerful way to tie ideas together. In the following activities, you are asked to consider unifying themes of this type.

Activities

4-6. Read about *Advance Organizers* on pages 171-172 in the Reader.

How can you teach the abstract concept of group before you teach the related, subsumed concepts? Study the following classroom activity for one possible approach; and then respond to the questions following the activity.

Middle School Level:

Have a student face the front of the room (back to the class) and label that position, **As You Were**. Then from that position, have the student respond to the commands: **Right Face, Left Face** and **About Face**. In each case, have the student return to face the front of the room before giving the next command. Be sure all agree with the results of the commands. Have other students try these out as a check. Then, work out the results of the student performing any two commands in sequence from the **As You Were** position. For example, a **Right Face** followed by a **Right Face** places the student in the position achieved by the single command, **About Face**. The questions being asked the students are: (1) Did the student respond correctly to the commands? and (2) Could we have used just one command to get the student in the same position from the starting point? Another student could be asked to stand in the **As You Were** position and respond to the single command suggested by members of the class. Ask if any other command would work just as well. Have them test their ideas by taking the floor and responding to the command. You will probably have to remind the class that the student must start from the **As You Were** position. After two or three more pairs of commands are tried, help the students design a table for the results. This is a good time to introduce symbols for the commands (*As, R, L, Ab*) and for the operation, **followed by** *(f)*. If the students begin to see a pattern, insist that they test it or defend the purported pattern. For example, some students will begin to realize that the order of the commands does not make a difference; others may not.

a. How does this lesson (the follow-up question/answer and summary section has been omitted) provide an introduction to the concept of mathematical group? What components of that concept would you want to address in your follow-up questions and comments?

b. Students at these grade levels are familiar with addition and multiplication tables. How can you use that familiarity to make some mathematical connections (likenesses and differences) between these calculation tables and that of a group?

c. How would you present the related subsumed concepts and principles (*e.g.*, set, binary operation, closure, Abelian or commutative group), given this lesson segment?

4-7. A visual approach or a laboratory activity can serve as a unifier that helps students retrieve concepts or principles.

a. Secondary students are taught to sketch the sine and cosine curves, showing only the intercepts, and maximum and minimum points. How would you suggest that they use the image of a sketch to retrieve information on period, amplitude, and value of the sine (cosine) of special angles?

b. How could you help students use the activity that follows as a way to retrieve information about various multiples, factors, and even divisibility rules?

Twenty-eight students in a middle school class are given number names, 1–28, in order across the rows. Then they are told to rise and stand in place as the teacher, Mr. Brant, calls out their number names. Mr. Brant calls out 2, 4, 6, ..., 28. Then he has the students sit and calls out a sequence of odd numbers. In each case, students are told to look at the spatial pattern. Multiples, primes, and other number patterns are used, with questions such as, "What comes next?" or "Why didn't I say that number?"

In this last activity, the teacher asks the students to look at and reflect on a spatial pattern formed by the group of standees (or by those still seated). It may be more or less difficult for them to see a pattern for certain sequences of numbers and for certain arrangements of students. For example, suppose the 28 students were seated as shown in Figure 4.1(a).

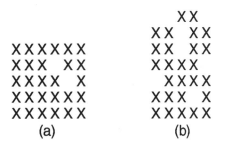

Figure 4.1: Two class arrangements of 28

The empty seats may present a problem for some students. However, the teacher can move students and/or follow-up with a hundreds chart of numbers in which there are no uncomfortable spaces. Contrast the seating arrangement in Figure 4.1(a) with that in Figure 4.1(b). Is one arrangement more likely than the other to help students recognize the spatial pattern that then leads to learning and retention of a number sequence? Will your answer vary with the number sequence presented? Try out a few examples and discuss the issue with another colleague.

Instructional strategies

In the last set of activities, you were given illustrations of two partial lessons that depended on the use of unifying themes or patterns. Earlier in this chapter, the notions of connections, integration, and unifiers were related to the concept of pattern. Looking for patterns reminds us of the process called inductive reasoning. In fact, a well-known mathematics educator, W.W. Sawyer, equated mathematics with patterns.

> *Mathematics is the classification and study of all possible patterns. Pattern ... is to be understood in a very wide sense, to cover almost any kind of regularity that can be recognized by the mind.* (Sawyer, 1955, p. 12)

Pattern is perhaps the most pervasive unifier to keep in mind as you plan daily lessons or outline units. As you read the textbook presentation, search for new ideas in journals or work out practice examples, you will want to be on the alert for the presence of patterns. What kind of patterns?

One kind might be labeled pattern by analogy. Reading about and thinking through an approach to a lesson, you may decide to compare a procedure in algebra to one from arithmetic. Multiplication of numbers in arithmetic is like, but not the same as, multiplication of algebraic expressions. What likeness can you exploit in a lesson? What are the differences that must be shared with students? The use of analogy is valuable, but must be undertaken with care. Let's consider what might happen if you teach students that adding $3x$ and $5x$ is like adding 3 apples and 5 apples and thus equals 8 apples or $8x$, whereas $3x + 5y$ is like the sum of 3 apples and 5 oranges and thus cannot be simplified further. Students may well question why $3x$ times $5y$ can be written as $15xy$. After all, they could argue, it doesn't make much sense to say that 3 apples times 5 oranges is the same as 15 apples oranges. This backlash from the use of analogy in teaching usually results from an approach in which the useful similarity is mistakenly treated as an absolute one-to-one correspondence. Not only must the teacher remember that an analogy is not a rule but a likeness, and thus never provides an exact fit, but the teacher must also convey that notion to the students. It's useful to remember the aphorism from logic that *every analogy limps.*

Does this problem mean that the use of analogies should be avoided? Absolutely not! Analogies are extremely useful ways to help students make an abstract process more concrete. However, analogies must be backed up with the mathematical rationale that undergirds the concept or procedure. Once again, the **Mathematical Connections** need to be stressed.

It's important to help students learn that a rule can always be derived from earlier mathematical rules or agreements. In this case, $3x + 5x$ can be rewritten as $(3 + 5)x$ either by thinking of the process as applying the distributive property, or by thinking of the process as factoring out a common monomial factor. Reflect on the connections between these two approaches. If the students had earlier been engaged in mental arithmetic lessons, either of these would seem reasonable extensions. In mental arithmetic lessons, the students might have been encouraged to learn how to rename sums to make the mental activity easier (*e.g.*, $36 + 48 = 12(3 + 4)$).

Sometimes the pattern is actually the rule itself. That was the case in the preceding example. The students would have to be helped to see that in the intermediate step, $3 (12) + 4 (12)$ followed the same pattern as $3x + 4x$. That is precisely what a formula or algorithm is— a patterned approach that can be used for all cases that fit. If you think that sounds a lot like type-problem solving and rule learning, you're correct. However, teachers must be sure that the pattern they taught is the one the students learned. Consider the common student error of reducing 24/134 to 2/13. The students have learned the pattern: *When the same numbers appear in the numerator* (they'd probably think "top") *and denominator of a fraction, cross out the numbers*. They would be correct, if the original expression had been:
$(2 \times 4)/(13 \times 4)$. (An illustration of one way to remedy this kind of situation can be found on page 61 of *Secondary Mathematics Instruction*. The example includes the use of geometric shapes to highlight patterns so that students must focus on the **shape pattern**, rather than on a mechanical procedure.)

How can we find out whether students have learned a misconception that has not yet shown up? It is possible for incorrect procedures to be hidden, you know. If the practice examples, by chance, can be correctly solved using the erroneous pattern, it may be some time before the teacher is aware that the student has learned, and perhaps overlearned, an incorrect rule. Let's reconsider the *crossing out rule*. When a student applies this technique to 16/64, that student gets the correct answer. As in earlier chapters, we return to the importance of ongoing diagnosis. When students keep a regular journal, they can be asked to write a new procedure in their own words and to show how it applies to examples **they** make up. Student pairs who are working together on a set of practice examples could be asked to agree on an explanation they would give to a new student who needed to be taught this procedure.

Some student misconceptions seem to occur so regularly that your cooperating teacher or other experienced teachers will alert you to their probable presence. Rather than identifying the incorrect procedure in words and warning the students against it, I would suggest that you check your set of in-class practice problems and be sure to include both some that can be solved correctly by the misconception and some that cannot. The pattern of errors will alert you to the presence of a problem before it becomes ingrained, and then you can decide whether individual or cooperative group efforts are the best way to solve the problem.

> **Hint:** Be sure to move slowly in the development of a lesson. Use concrete and visual examples. Don't force students to verbalize a rule in final form too soon. If they are succeeding by using a precursor algorithm, praise them for their ability. If you want them to move toward the final, shorter form of the algorithm or formula, assign examples that are not so conveniently solved by the precursor. Be sure to emphasize the pattern form that links new complex-looking examples with early easy examples. Ask students to state new procedures in their own words and allow them to return to concrete materials, when necessary, to solve a problem. Carefully plan the sequence of illustrative examples you intend to use to develop a new rule. Most of us unintentionally build in unwanted patterns when we give students examples extemporaneously. Such unwanted patterns may lead students to identify and learn an erroneous rule. And, finally, students may create a new rule that is equivalent to the one you had in mind. Praise them for finding this mathematical connection and work with them to demonstrate the equivalence. If it works efficiently, don't insist that they abandon it.

Ongoing diagnosis is an important aspect of instruction. Sometimes the teacher has prior knowledge of likely misconceptions and can design probing questions, practice examples, journal prompts, or follow-up lab activities to identify the existence of these misconceptions. Even when the teacher has no specific prior knowledge of likely trouble spots, well-designed feedback strategies that correspond to the kind of mathematical content in the lesson are essential. In a concept plan, there need to be feedback strategies designed to yield data on whether the students can generalize the concept to new instances. In a rule plan, there must be opportunity for the students to demonstrate that they can apply the rule to new instances. (Examples of getting feedback in sample daily plans are analyzed in Chapter 7 of *Secondary Mathematics Instruction*.) It's important to reflect on the two-fold purpose of these strategies. They assist the teacher in making on-the-spot judgments as to what should happen next in class, and they provide data on the extent to which the students have understood the mathematical connections in the lesson. However, as important as feedback-getting strategies are to the identification of missed or erroneous connections made by the students, feedback-giving strategies are equally important in emphasizing, or redirecting student attention to, mathematical connections.

Activities

4-8. The following questions are based on the probability lesson found on pages 145-146 of the Reader.

a. When the teacher notices that three groups are confused about the event using the roulette wheel and the spinner, how does the teacher give feedback that helps students to find connections?

b. Saul and Alice have a disagreement about the way to treat the probability of two events happening at the same time. What type of feedback is given by the teacher and how does that feedback appear to highlight mathematical connections?

4-9. Refer to the suggestions for homework post-mortem found on page 173 of the Reader. Choose any one of the suggested approaches and study it for the feedback-getting and feedback-giving strategies used by the teacher. With a colleague, talk about the way(s) in which partially or completely missed mathematical connections could be identified, reinforced, or redirected when those feedback strategies are used effectively.

In Activity 2-5, similar questions were asked about the feedback strategies in the probability lesson. However, those questions referred to communication; these in Activity 4-8 refer to connections. Examine both sets of questions and reflect on your responses in each case. To what extent could these feedback strategies be considered as ways to integrate instruction?

With effective use of appropriate feedback strategies on a daily basis, it might seem as if the need for getting feedback in the long-term is negligible. That would be true if mathematical connections were merely horizontal or vertical links. However, we know that mathematics is more complex than that. If we were diagramming any particular area of mathematics, there would be chains. Some of these groups of chains might be related in a hierarchical fashion. Others might form clusters related in a superset/subset fashion. In teaching, we're continually merging the small chains into larger ideas. Thus, it is important to stop at critical points during a unit and include a feedback-getting task to find out if students have been able to make these overarching connections. Journal writing may be used as one diagnostic tool, provided that the probes are specifically stated. Two examples are given here.

> Write about your understanding of the relationship between congruent figures and similar figures.

Devise a chart, with an explanation, to show important characteristics of special numbers, such as $\sqrt{-1}$, e, π, 0, google, $\sqrt{2}$.

Sometimes a sum-up study or practice session can serve the long-term feedback-getting purpose. Small groups can work cooperatively on a specially designed handout, a This-Could-Be-Your-Test set of problems, or a set of problems where each is accompanied by a complete but wrong student response. In the latter example, the imaginary student may have gotten the correct final answer by mixed, wrong procedures, one of which cancelled out the effect of another. The small groups are asked to identify the error(s), correct them, and decide why the student made that error (*i.e.* what connections the student missed) and how to correct that student's thinking. Compare this suggestion to that in Activity 1-9. In Activity 1-9, the emphasis was on novel-problem-solving. Here the emphasis is on student self-evaluation. Thus, the kind and scope of the problems on the sheet might differ. In particular, if the primary intent were student self-evaluation, the teacher would include problem context in which there is some potential for missed connections.

Another feedback-getting tool used with success by some teachers is the concept map (*webbing* or *spider graph*). Study the student concept map on page 174 of the Reader. This seventh grade student was asked to link a list of given concepts after instruction on those concepts. Notice that the student map differs in at least one major respect from the teacher map referred to in Activity 4-3. The student is trying to portray what the different concepts mean in terms of each other, as the student sees it. In most cases, you can read the rules that link one concept to another. The teacher, in contrast, was not restricted to using definitional links but could identify any functional or structural link between ideas. These differences help to remind us that the connections we are aware of as teachers may not be the ones students learn. When these connections are important to learning of future mathematics, it is important that we assess student understandings of them. Concept maps are one excellent way of beginning this process.

Beginning this process is the phrase used here. Whether the teacher uses journal entries, a sum-up study or practice session, concept maps, selective questions, or any other tool, these are all only beginnings. Follow-up questioning of individuals and/or the entire class and clarification in the form of demonstrations, mini-lectures, model solutions on acetate, handouts, or computer screen feedback are necessary next steps.

We need a way of summarizing the nature of the instructional strategies focused on mathematical connections—a pedagogical unifier. One that you may have read about before is a set of descriptions of the acts that should occur in any lesson. The set is called *The Events of Instruction*. (You'll find them listed and briefly described on page 41 of *Secondary Mathematics Instruction*.) Appropriately enough, this description is in the chapter on feedback, and we already have had many examples of the importance of feedback strategies to instruction on mathematical connections. The table that follows was designed to give illustrations of a strategy (or strategies) corresponding to each of the events. In order to make the connections as explicit as possible, the second column refers to events in the middle school lesson on the concept of π found in the Reader on pages 153-154. It's important to note that this concept lesson would precede a lesson on the formula, $C = \pi d$. (For a sample plan on the formula see pages 170–172 of *Secondary Mathematics Instruction*.)

Table 4.1: Sample Strategies Matched to the Events of Instruction

Events of Instruction	Applied to a Lesson on the Concept of π
Gaining and controlling attention Feedback on what to observe; relevance to past work	Teacher shows a variety of cans and rectangular boxes and poses problem of calculating the distance around. Question and answer follows: how to do this for boxes, relationship of various sides to the perimeter.
Informing student of expected outcomes What's expected of students	Teacher demonstrates use of string for measuring and questions its convenience. Teacher tells students to look for a pattern.
Stimulating recall of prerequisites	Question and answer session regarding labels for circumference and diameter, and how to measure each.
Presenting new material	Students work on measuring lab, record data on provided worksheets, calculate C/d.
Guiding the new material	Teacher helps with measuring troubles and identifying nature of pattern. Question and answers on reason for calculation differences raises issue of proof. Teacher gives mini-lecture to sum up extent of match of lab results to value of C/d, related history, and label of π. Teacher begins to move toward use.
Providing feedback	Teacher does so throughout the lab and during question and answer about results. Praises, helps, cues.
Appraising performance	Teacher moves among groups, observing, and cueing. Question and answer on validity of final result. Teacher provides follow-up question and answer to see if students can generalize.

The first two *events of instruction* might be thought of as helping students to know (1) where they're headed today and why it's important to get there, and (2) what that journey has in common with past mathematical excursions. The teacher provides the students with a rationale for the lesson, but notice how that rationale must be framed from the students' perspective, not the teacher's. Thus, the concept label is not provided until the students have some understanding of the concept itself. Later when the label is introduced by the teacher, another rationale is provided for the label—a brief historical background, which is itself another mathematical connection.

Let's assume that the π lesson concluded with a handout to be started in class and completed for homework. This homework sheet is reproduced in Figure 4.2. Study these questions to ascertain how they could be used to appraise performance of the students after, or at the end of, the lesson on the concept of π.

π **Lesson Sheet**

1. Astronomers measured the circumference and diameter of Mars. Geologists measured the circumference and diameter of Earth. How would you expect the quotient of C/d to compare for Earth and Mars? Why?

2. The diameter of a hula hoop measures 80 cm. The circumference of the hula hoop is unknown. Mary claims she can still tell you exactly what C/d equals. Jim says, "If measures are taken, then C/d won't come out a constant." Who is right and why?

3. Group 7 in Ms. Mason's class handed in this lab sheet. The leader, Jerry Bart, complained that some joker had erased part of their work. Help Jerry by filling in the blanks.

Object	Circumference	Diameter	C/d
Soup can	21. __ cm	6.9 cm	3.1
Jug	__ .0 cm	10.0 cm	3.1
Half dollar	__ . __ cm	4.0 cm	3.1
Disc (large)	62.0 cm	__ . __ cm	3.0

Figure 4.2: Follow-up assessment questions

Activities

4-10. Take a few minutes now and, with a colleague, identify the other places in the lesson referred to in Table **4.1** where the teacher is alerting students to mathematical connections. What are these mathematical connections?

4-11. Analyze the assessment sheet in Figure **4.2** and, with a colleague, identify the mathematical connections the teacher is trying to reinforce.

In your response to Activity 4-10, did you identify the teacher's attempt to reinforce the abstract nature of mathematics through the questions on the measurement process? The teacher went even further and emphasized that measurement labs can't lead to mathematical proof. Go back to page 153 in the Reader and read the section in which Mr. Cain tells the students what numerical approximations they will use. What do you think of the way in which he introduced the number 22/7? Was a connection made? Not really. That number came out of nowhere as far as the students were concerned. Is it necessary to introduce the value of 22/7 at this stage? It doesn't seem so from the questions on the assessment sheet. Why do textbooks always include this number in lessons on π? This may be a good time to check a historical source on approximations of π to find out when this particular fraction became associated with the irrational number. An excellent source for your search, and one that should be owned by every teacher of mathematics, is *Historical Topics for the Mathematics Classroom*, recently reprinted by the National Council of Teachers of Mathematics (NCTM, 1989b).

Through the medium of the *Events of Instruction* , we have seen that strategies emphasizing mathematical connections link ideas in meaningful ways, reflect the structure of the subject, and show students a glimpse of the human inventiveness that helped to develop mathematics. Equally important for all of the same reasons are strategies emphasizing the connections between mathematics and the world outside mathematics, in particular the everyday world of the students. This is the area to which we now turn our attention.

Connections Outside Mathematics

If a teacher taught mathematics in such a way as to emphasize all possible connections within mathematics, but did not attend to the connections between mathematics and the real world, that teacher would still not have captured the spirit of the *Standards*. Moreover, with the exception of those students who view mathematics as a kind of game of mental skill, others would lose interest and question the use of all these procedures and abstractions. Perhaps, most important, none of the students would have learned an important aspect of the structure of mathematics, the role of mathematical model.

An important source for this section on **Mathematical Connections** is Chapter 10 in *Secondary Mathematics Instruction*. The activities and background in that chapter represent a blueprint for the teaching of connections between mathematics and the everyday world. The illustrations include complete lessons in which the interface of mathematics and some other subject are treated. The set of honeycomb lessons (pages 158-160 in the Reader) is an excellent example of this kind of detailed approach to the teaching of mathematics and science concepts. The illustrations also include ideas that might be used to get attention on a topic, to enliven practice examples, to start a discussion, or to reinforce a concept. Cartoons, advertisements, toy cars, soup cans, basketballs—all of these materials and objects when used in conjunction with a well-framed question teach mathematics through the everyday world. The basketball may be used in a demonstration of spherical objects. The soup can may be passed around the room so that students can identify the kinds of numbers on the label. In either case, the real-world object has the potential for getting attention on the mathematical concepts to be considered. Thus, just as in the case of connections within mathematics, connections outside mathematics may be introduced at specific points in a lesson or may actually infuse the entire lesson.

One of the best ways to prepare for teaching connections between mathematics and the real world is to build a resource file of ideas for each course you teach. (Directions for developing a file can be found on pages 342–344 of *Secondary Mathematics Instruction*.) Ideas can be filed on cards or sheets in a course resource file. For many teachers, it's easier to locate the ideas if they're filed under a unit heading. Some ideas will be transferred from mathematics methods texts. Additional ideas can be located in source books (see the suggestions at the end of Chapters 10 and 11 of *Secondary Mathematics Instruction*) and teaching journals. The School Science and Mathematics Association recently rededicated itself to fostering the integration of science and mathematics. Its journal, *School Science and Mathematics*, regularly contains a column with a lesson integrating science and mathematics, as well as individual articles in each subject. The journals of the National Council of Teachers of Mathematics include sample lessons and teaching ideas, many of which involve connections with the everyday world.

However, often the ideas that work the best are the result of your adaptation of clippings, cartoons, a piece of geometrically designed wallpaper, a photo cube, data from an article on teen-age drinking, or any other attention-catching object. Perhaps the fact that you've brought these objects from their customary "residences" to the mathematics classroom has something to do with their effect on the students. How do you decide what objects to save and how they should be used?

You need to do some background work again. Actually, if you've done the careful analysis of subject matter described in an earlier section, one part of your background work is almost complete. The other part of the subject-matter preparation is to analyze real-world connections for each major topic area and concept. For example, if you were preparing a lesson on the parabola, you would try to collect examples of real-world models—perhaps the path of water in the school drinking fountain. It's important to reflect on the real-world phenomenon and its connection to the mathematical concept. Here are some questions you might consider.

> To what extent are the water path and the parabolic curve alike? How are they different?

> Would it be possible to construct a quadratic equation that would be a good predictor for the water path?

> What measurements would you take?

If you do that kind of reflection ahead of time, you're more likely to be attuned to potential connections between other real-world phenomena and mathematics.

Activities

4-12. For each of the mathematical concepts given, describe at least one real-world connection (*e.g.*, physical model, real-world illustration): decimal number, equivalent fractions, function, probability, volume.

4-13. You are given a box containing the following objects: a carpenter's ruler, paper clips, two mirrors, plumber's tape, buttons of various sizes, drinking straws, toothpicks, mini-marshmallows. Choose any one or more of the objects and explain how you would use them to show a real-world connection to some aspect of mathematics.

Each of these activities has a variety of possible responses. The dial on a gasoline pump or the odometer in a car uses decimal numbers. After the number of decimal places has been identified, you can ask the students if they know why the dials are graduated to that decimal place. Try representing equivalent fractions by using egg cartons and candy corn, but be careful. If you fill 6 cups out of 12 (6/12) and then fill 3 cups out of 6 (3/6), the students aren't likely to view the two arrays as equivalent (Figure 4.3).

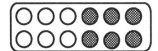

Figure 4.3: Representing fractions

That mistake reminds us that it is only meaningful to call two fractions equivalent if they are fractional parts of the same whole. Back to the importance of the real world!

If you're wondering what on earth to do with the mini-marshmallows in Activity 4-13, think back to science class. Science teachers use them a lot as joints in models. With that cue, you should be able to create a variety of geometric figures with them and the toothpicks. Paper clips are a nice example of a simple curve that is not closed. They also can be chained together to serve as a measuring instrument. Does a paper clip possess symmetry? What kind of equation would yield a paper-clip curve? Share your ideas on these questions and on the other responses you had for both activities with a colleague.

Along with this kind of mind-stretching, you need to scan the funnies, the sports pages, editorials, and economic and political graphs for their classroom potential. Cartoons that use strange logic, refer to mathematical concepts or processes, or actually picture a mathematical concept (*e.g.*, a mirror image or a symmetrical relationship) are wonderful material for interesting homework assignments, classroom discussions, and even tests. Sets of statistical data should be examined for their possible use in mathematics class. Finally, you'll want to collect information on the interests of your students.

It's true that we can make some generalizations about the interests of a particular age group. For example, secondary school students are concerned about acceptance by peers, and thus the ways in which they might seem different from peers—their clothes, their hobbies and after-school activities, extremes in height or weight, *etc.* (For other ideas, see pages 191–192 in *Secondary Mathematics Instruction*.) Those ideas are still good starting points in your efforts to build a resource file of real-world connections. However, it also helps to get specific data on the interests of the individual students in your class. One way to do that is to structure a writing assignment in which the student is asked to describe herself/himself. The assignment can be given on the first day or two of the year with the teacher explaining that student interests will be considered as lessons are planned. You can suggest areas of interest, such as hobbies; summer or part-time jobs; favorite music, TV shows, or movies; and participation in sports, band, chorus, or other school clubs. By making the assignment open-ended, you will not get information on each area of interest, but the choice will leave the student in control. That's a preferable way of ensuring that the information you do receive is what the student wants you to know. Writing was considered in Chapter 2 as one of the important ways of communicating mathematics. Refer to that chapter for ways to ensure that the students aren't threatened by a writing assignment.

> Hint: One of the suggestions given in Chapter 2 was for the teacher to write while the students are writing. That will only work if the teacher shares that written work with the class before the period ends. How does the teacher balance the need to share something personal with the class with the problem of sending mixed signals on the teacher/student relationship? You might include some of your hobbies or avocations, your hopes for the class, *etc.*. A good guideline for the teacher in the writing assignment is to share background that might help the students appreciate some of the strengths the teacher will bring to class. *I enjoy hiking, birding, and outdoor activity. I'm always looking for mathematical patterns in growing things—for example, the spirals on pine cones.* That kind of sharing of your interests makes sense.

> If you are student-teaching and want to try this assignment, talk over your ideas with your cooperating teacher. You'll want to be careful not to appear to be pleading for attention by becoming *one of the kids*. They know you're not one of them, and if they think you're confused about the role of teacher, the class atmosphere will soon deteriorate. (It might be helpful to read "So You're A Teacher" on pages 336–338 in *Secondary Mathematics Instruction*, just to clarify your own thinking.)

Instructional strategies

In order to help students meet the outcomes listed under Standard 4, the teacher must design interface lessons on a regular basis. However, such lessons should be scheduled so that they make sense in terms of the previous day's work, are followed up the day after the lesson, and are referred to in summary sessions and on quizzes and tests. Some of the interface lessons mentioned in previous chapters in this text include *The Hobbit* lesson, the honeycomb lesson, the sunflower seed lab, the bicycle gear lesson, the traffic count lesson, the lesson on human growth patterns, and the Chambered Nautilus lesson. Each of these lessons involve regularly taught mathematics concepts and procedures. In order not to lose the value of the lesson, the interface concepts and the references to the concrete materials or real-world phenomena should appear in later homework and quiz questions. The introductions to new topics might begin by showing students a model used in one of these lessons—*e.g.*, a sample honeycomb model—with recall questions on the results of that lesson and follow-up questions on a related topic, such as surface area/volume relationships.

Another example of an interface lesson is based on an idea from *Global Atlas* (Draper, 1991). The charts, graphs, and maps in *Global Atlas* can be used to show connections between mathematics and environmental, political, and social issues. The introductory material includes suggested ways of studying photographs, diagrams and statistical tables. In fact, the authors of this source use the phrases *idea mapping* or *webbing* to describe a way to organize investigations. Sound familiar? The term *webbing* has been used earlier in this text as a broad synonym for *concept mapping.* The authors of *Global Atlas* provide key questions and appropriate information in pictorial, graphic, and tabular form to help guide investigations. For example, the starter exercises for the topic "The Earth in Space" are:

How does the Earth compare to other planets in the solar system?

Describe the movement of the Earth during the course of one year.

Depending on the level of the class, the teacher might add second-level, more specific questions. Students could graph the distance between the planets and the Sun, study the shape of the orbits, and be led into an investigation of the historical belief in the circular shape of the orbits. "Wow! Neptune is 4,502.7 million km from the Sun!" Graphing distances that range from 58.3 million km to 5894.2 million km seems to be a formidable mathematical problem. Students could be asked how they might rewrite each distance to make the numbers more workable. One possible suggestion is to rewrite each as 10 times a decimal and then round the decimal to the nearest whole number. These results will still require some clever scaling of the axes, if the students were restricted to the usual 8 by 11 sheet of graph paper. Why restrict them? If larger sheets are not available, allow the students to tape sheets together and work at a table, window ledge, or even the floor. Now a fruitful discussion on the validity of the transformations of data, the meaning of the transformed data, and the construction and interpretation of the graph may take place.

When students study the shape of the orbits of the planets, connections can be made to events such as eclipses, space travel, and the seasons in various locations on Earth. The students can be asked to make conjectures on the effect of the planets' distances from the Sun and, with the cooperation of the science teacher, research some of the available information from recent space travel and studies of the solar system. An examination of the orbits of the planets provides a unique opportunity to learn about the ways in which pseudo-science, philosophy, and religion interacted to affect scholars who were studying the planets. This kind of student investigation was suggested in Chapter 3. An excellent source for an approach that uses explanations based on mathematics accessible to secondary school students can be found in Polya (1963).

Activities

4-14. The following data giving the diameter of the planets in miles are from Bell (1972).

Earth	Jupiter	Mars	Mercury	Neptune
7920	88640	4200	3100	31000

Pluto	Saturn	Sun	Uranus	Venus
2110*	74500	864000	32000	7700

* In 1972, Bell was not able to list the diameter of Pluto since scientists had not been able to gather sufficient information to calculate this measure. More recent advances in science have provided us with these data.

a. Transform the data from miles to km.

b. Arrange the planets in order from smallest diameter to largest, and plot the diameter against the distance from the Sun. Be sure the data on each axis are transformed in the same way. (Suggestion: On the horizontal axis, draw circles for each diameter so that the relative size of each planet can be visualized.)

c. Identify the connections involved in a lesson based on this graphing activity. What mathematical concepts and procedures would be reinforced?

In Activity 4-14, relationships among the planets are explored further by considering one aspect of size, and graphing that against distance from the Sun. Very large and very small dimensions are enormously difficult to comprehend. In this activity, by transforming the data and picturing the cross-section of the planets on the horizontal axis, students have an opportunity to make sense of relationships among and between planets. You would still need to work on the concept of the actual size of the planets. It may help to consider Mercury, which has a diameter of 3100 miles. The distance across the continental United States is approximately 3000 miles. Have your students imagine a belt across the country and then the cross-section of a sphere, with that belt as diameter. This kind of imagery of distances that may be known experientially to some students helps to make a difficult concept of size concrete.

Interface lessons provide an opportunity to explore the concept of mathematical model and the reasons why one model may be discarded in favor of another in view of real-world needs. In Chapter 1, you were given the example of the engineer who accepts and uses mathematical models, knowing that there will be some differences between these and the corresponding real-world phenomena. Think of the engineer's formulas as theoretical models. In earlier eras, these relationships might well have been tried-and-tested rules. Before computers, scale models of a structure would be built on the basis of the formulas. The scale model would be subjected to scaled-down stresses proportional to the stresses expected in the real world. A relatively large margin of error, in favor of safety, might well have been built into the actual structure after a sufficient number of tests had been conducted. Computers have allowed engineers to build computer models, to simulate stresses to greater precision than ever before, and thus to test the applicability of the theoretical relationships more extensively than before. Once again a margin of error is built in, but it is likely to be a smaller margin than before, due to the extensive evidence provided by computer simulations. Can we ever say with confidence that technology has provided us with a perfect model? No, we can't, because a mathematical model is just that—precise and perfect for an abstract world, whereas the real-world phenomena are irregular and never able to be measured **exactly** (with the exception of counting), even if the surfaces appear perfectly flat, or the corners perfectly sharp.

One way to introduce this aspect of mathematical model into the classroom is to invite a local engineer, perhaps a student's mother, to visit class and talk about this part of her work. Human resources may be used in many ways to make connections for the students. For another example, a father who is also a nurse might describe some of the mathematics he uses when preparing dosages for patients. He may also be on a medical diagnostic team, whose members input patient data into computer programs. Such programs may provide the team with a diagnostic plan, or a patient glyph. The latter is a "picture" of the patient's relevant health conditions.

Another human resource who can help you enhance interface lessons is a colleague in the related discipline. There is no doubt that scheduling in some schools is not conducive to this kind of team, or cooperative, teaching. However, you may be able to begin to work toward this ideal by talking with a colleague about your plans, asking for information on related issues (connections), and suggesting cooperative approaches. (Some suggestions for collaboration with the science teacher are provided on pages 280–282 of *Secondary Mathematics Instruction*.) Social studies teachers would be valuable resource persons for some of the lessons based on data in *Global Atlas*. How could the English teacher help you deal with long-term projects or research papers? If you can't think of a way to involve the physical education teacher, refer to the muscle fatigue lesson found in *Secondary Mathematics Instruction* on pages

278–280. Students could also consult the science teacher and school nurse for extension activities related to this lesson.

Are ideas emphasizing connections outside mathematics found only in interface lessons? You already know the answer to that. Remember the π lesson that began with the teacher holding up various real-world objects that had at least one circular cross-section. Earlier in this chapter, mention was made of starting a class with a cartoon related to symbolic logic or transformation geometry. The ideal, then, is to incorporate some form of real-world connection into every lesson. Often these are based on the students' here-and-now interests. For example, in a statistics unit, those on the soccer team can be asked to get data, or *stats*, on the performance of the players (with the permission of the coach). The players' names can be deleted. Class members can work on an analysis of the data to answer the kinds of questions asked by the coach during the season. Students who are artists at heart can be resource persons on what really is done when artists mix paints. They will also be helpful in topics on perspective drawing, the use of parallels, and the grid system used to enlarge a painting. Starting a lesson by having a band student create a little music on a clarinet is sure to get attention. When that experience is tied to concepts of pitch, frequency, and the relationship between length of the air column and pitch, the students' attention is maintained and an effective lesson is ensured.

All of these examples and the wealth of others in this text provide us with a key to the contribution of real-world ideas to a lesson. Table 4.2 is designed to highlight these contributions.

The importance of real-world strategies for getting and keeping attention cannot be overstated. Why would any of the strategies in Table 4.2 have potential for motivating the students to learn the corresponding mathematics? Why do real-world strategies make sense in terms of our understanding of cognitive development theory? Yes, even for apparently formal operational youngsters, lessons that build in connections to the real world capture attention and reinforce the concrete base so vital to the understanding of abstract ideas.

As we have found with the first three standards, lessons that include attention to outcomes under **Mathematical Connections** are likely to include strategies related to **Mathematics as Problem Solving**, **Mathematics as Communication**, and **Mathematics as Reasoning.** In the next chapter, we take an intensive look at the **Evaluation Strategies** related to these four curricular strategies. Before you turn to Chapter 5, take a few minutes and consider what that might mean for the evaluation of the outcomes we've studied so far. If you decided there would have to be striking differences from traditional evaluation, you're on target.

Table 4.2: Roles of Real-World Connections in Lessons

Role	Example
1. Getting attention of most students at any point in a lesson	Have a working clock displayed so students can see moving gears for a lesson on ratio
2. Keeping attention of students	Design practice problems with a storyline based on school teams
3. Providing a rationale for concept or principle being taught	Show and tell uses of conic light reflectors in stage design to get various lighting effects
4. Illustrating the special nature of mathematical models	Have students design a school recycling plan—where to place bins, how often to empty them
5. Helping students move toward novel problem solving	Create a quiz problem around a recent school event or the students' favorite TV show
6. Providing for individual differences	Have student stamp collectors show and talk about stamps with mathematical designs or mathematical history themes

References

Bell, M.S. 1972. *Mathematical uses and models in our everyday world.* Studies in Mathematics, vol. 20. Stanford, CA: School Mathematics Study Group.

Draper, G. 1991. *Global atlas*. Toronto: Gage Educational Publishing Co.

Farrell, M.A, and Farmer, W. A. 1988. *Secondary mathematics instruction: An integrated approach.* Dedham, MA: Janson Publications, Inc.

National Council of Teachers of Mathematics. 1989a. *Curriculum and evaluation standards for school mathematics.* Reston, VA: NCTM.

National Council of Teachers of Mathematics. 1989b. *Historical topics for the mathematics classroom*. Reston, VA: NCTM.

Polya, G. 1963. *Mathematical methods in science* Studies in Mathematics, vol 11. Stanford, CA: School Mathematics Study Group.

Sawyer, W. W. 1955. *Prelude to mathematics.* Baltimore, MD: Penguin.

Evaluation Standards for School Mathematics

STANDARD 1: Alignment

Methods and tasks for assessing students' learning should be aligned with the curriculum's

— goals, objectives, and mathematical content;

— relative emphases given to various topics and processes and their relationships;

— instructional approaches and activities, including the use of calculators, computers, and manipulatives.

STANDARD 2: Multiple Sources of Information

Decisions concerning students' learning should be made on the basis of a convergence of information obtained from a variety of sources. These sources should encompass tasks that

— demand different kinds of mathematical thinking;

— present the same mathematical concept or procedure in different contexts, formats, and problem situations.

STANDARD 3: Appropriate Assessment Methods and Uses

Assessment methods and instruments should be selected on the basis of

— the type of information sought;

— the use to which the information will be put;

— the developmental level and maturity of the student.

Reprinted with permission of the publisher. See acknowledgements.

STANDARD 4: Mathematical Power

The assessment of students' mathematical knowledge should yield information about their

— *ability to apply their knowledge to solve problems within mathematics and other disciplines;*

— *ability to use mathematical language to communicate ideas;*

— *ability to reason and analyze;*

— *knowledge and understanding of concepts and procedures;*

— *disposition toward mathematics;*

— *understanding of the nature of mathematics;*

— *integration of these aspects of mathematical knowledge.*

STANDARD 5: Problem Solving

The assessment of students' ability to use mathematics in solving problems should provide evidence that they can

— *formulate problems;*

— *apply a variety of strategies to solve problems;*

— *solve problems;*

— *verify and interpret results;*

— *generalize solutions.*

STANDARD 6: Communication

The assessment of students' ability to communicate mathematics should provide evidence that they can

— *express mathematical ideas by speaking, writing, demonstrating, and depicting them visually;*

— *understand, interpret, and evaluate mathematical ideas that are presented in written, oral, or visual forms;*

— *use mathematical vocabulary, notation, and structure, to represent ideas, describe relationships, and model situations.*

STANDARD 7: Reasoning

The assessment of students' ability to reason mathematically should provide evidence that they can

— *use inductive reasoning to recognize patterns and form conjectures;*

— *use reasoning to develop plausible arguments for mathematical statements;*

— *use proportional and spatial reasoning to solve problems;*

— *use deductive reasoning to verify conclusions, judge the validity of arguments, and construct valid arguments;*

— *analyze situations to determine common properties and structures;*

— *appreciate the axiomatic nature of mathematics.*

STANDARD 8: Mathematical Concepts

The assessment of students' knowledge and understanding of mathematical concepts should provide evidence that they can

— *label, verbalize, and define concepts;*

— *identify and generate examples and nonexamples;*

— *use models, diagrams, and symbols to represent concepts;*

— *translate from one mode of representation to another;*

— *recognize the various meanings and interpretations of concepts;*

— *identify properties of a given concept and recognize conditions that determine a particular concept;*

— *compare and contrast concepts.*

In addition, assessment should provide evidence of the extent to which students have integrated their knowledge of various concepts.

STANDARD 9: Mathematical Procedures

The assessment of students' knowledge of procedures should provide evidence that they can

— *recognize when a procedure is appropriate;*
— *give reasons for the steps in a procedure;*
— *reliably and efficiently execute procedures;*
— *verify the results of procedures empirically (e.g. using models) or analytically;*
— *recognize correct and incorrect procedures;*
— *generate new procedures and extend or modify familiar ones;*
— *appreciate the nature and role of procedures in mathematics.*

STANDARD 10: Mathematical Disposition

The assessment of students' mathematical disposition should seek information about their

— *confidence in using mathematics to solve problems, to communicate ideas, and to reason;*
— *flexibility in exploring mathematical ideas and trying alternative methods in solving problems;*
— *willingness to persevere in mathematical tasks;*
— *interest, curiosity, and inventiveness in doing mathematics;*
— *inclination to monitor and reflect on their own thinking and performance;*
— *valuing of the application of mathematics to situations arising in other disciplines and everyday experiences;*
— *appreciation of the role of mathematics in our culture and its value as a tool and as a language.*

Chapter Five

Evaluation Standards

In much of the curricular reform of the twentieth century, little attention has been paid to the importance of changing evaluation to correspond to the new curricula. In fact, a common complaint of teachers, when questioned about the lack of change in their classrooms, has been that they must teach *to the test*. The *test* refers to school-level exams, state exams, or broader standardized tests. To an extent, they are justified in their complaint. It's very difficult to expect children and parents to understand the importance of group work, communication skills, labs, and real-world projects if the assessment measures are restricted to multiple-choice tests of definitions and algorithms.

The NCTM authors were aware of this problem and incorporated a major section on evaluation into their document. In this section they faced the problems of student and program assessment. In this text, we will restrict our attention to standards describing the nature of evaluation—Evaluation Standards 1, 2, and 3—and to the student assessment standards—Evaluation Standards 4 through 10.

The Vision: *Beyond Tests*

Did you think the subtitle of this section should have been "Beyond Typical Tests"? I wouldn't be surprised at that reaction. Evaluation has been considered in the preceding chapters of this text and some of the assessment methods have certainly been atypical. However, the subtitle, "Beyond Tests" was chosen to reflect the NCTM authors' concern that too many people consider evaluation to be a synonym for tests with accompanying grades. They wanted to ensure a much broader view, encompassing feedback-getting strategies as well as the kinds of assessments that lead to grades. Did you notice that these standards were not sub-divided by grade grouping? In the *Standards*, illustrative examples for grades K–12 are included under each of the student assessment standards. However, our attention will be restricted to the examples for grades 5–8 and 9–12 as was the case in the first four chapters of this text.

In the first three evaluation standards, the authors emphasized basic assumptions endorsed by all those who study and work in the field of evaluation. Why is it necessary to include these assumptions here? For example, every standardized test manual includes consideration of **Alignment**. Usually a table of specifications is provided to show the relative emphases given particular bits of content and the levels of difficulty at which they will be tested. However, unlike traditional content, the content in the *Standards* includes both process and product, and both must be evaluated. Thus, the outcomes under the Alignment standard refer to process and product, and to the approaches and materials used in instruction. For example, if the students have had pattern-seeking experiences with calculator-based examples, evaluation measures should not only assess their ability to identify patterns but include some items that are calculator-based.

In the second evaluation standard, **Multiple Sources of Information**, the authors highlighted the complexity of evaluation and the need to sample understanding frequently and in different ways. For instance, a particular outcome might be assessed by collecting observational data on a scale or in anecdotal reports, by reviewing the results of a computer project, by collaborative grading of a research paper by the mathematics and science teachers, or by scoring a traditional paper-and-pencil test. Again, this is an accepted evaluation principle. However, the interpretation of the principle differs from usual practice. Under the best of circumstances, classroom teachers in the past have interpreted this principle by assessing the same outcome with questions on several short quizzes and on a unit test. Within the context of the *Standards*, outcomes should not only be assessed on more than one occasion, but with quite different types of measures.

In the third evaluation standard, **Appropriate Assessment Methods and Uses**, the authors delineated three kinds of assessment from the student point of view. The NCTM authors classified these as diagnostic, instructional feedback, and grading. An examination of Table 3.1 in the *Standards* shows that written tests are one method that can be used not only for grading purposes, but also to provide instructional feedback. It's easy to get confused about the authors' intent here. Let's think about the relationship of the three types of assessment in terms of diagnostic, formative, and summative evaluation. These three kinds of evaluation do sometimes overlap, as depicted in the example given on page 206 of *Secondary Mathematics Instruction*. The figure is reproduced here (Figure 5.1).

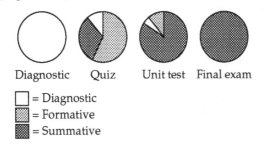

Figure 5.1: Relative emphasis of evaluation

Notice that only the first and the last circle depict a unique evaluative use. The other circles show a single instrument being used for a grade (summative evaluation) and for instructional feedback (formative evaluation). There's also a tiny sector labeled "diagnostic evaluation." How can such tests be used for diagnostic evaluation? According to Farrell and Farmer (1988, page 205), the two purposes of diagnostic evaluation are

(1) to place a student at the proper instructional starting point

or

(2) to find out the causes of instructional defects that have been isolated during instruction.

Despite this apparent use of a single assessment for all three kinds of evaluation, it is still necessary to reinforce the importance of designing assessments primarily for one of these purposes.

For example, if prior to instruction on problem-solving you wish to diagnose student learning of the prerequisite skills, you could design a scale, and rate the students' tactics as they play a computer game. However, you would not include these ratings among the grades on the students' achievement measures. Similarly, a structured tour in which the teacher observes progress of individual students using a new algorithm is an assessment that is primarily formative. The teacher is getting feedback on the extent to which the students have understood the first part of the lesson. An example of that can be found in all lessons aimed at presenting a new rule or algorithm. You've read about this kind of assessment in Chapter 1 of this text, when strategies for type-problem lessons were outlined. The teacher tours during supervised practice, getting feedback from, and giving feedback to, the students. At the same time, the teacher's plan had to include alternative next steps, depending on the progress of the students. No grade is associated with student performance on this formative assessment. However, the teacher will subsequently design a task that is primarily summative in nature—one that will assess the students' knowledge of the algorithm. The implications of the third evaluation standard affect day-to-day planning and instruction. We will return to some of those implications in the next section.

Evaluation Standards 4–10 apply to student assessment. There are three standards (5, 6, and 7) labeled to correspond to three of the four unifying themes treated in this text. At first glance, it seems that there is no evaluation standard corresponding to Curriculum Standard 4: **Mathematical Connections**. Are the outcomes under this standard not to be assessed? That just can't be the case. Take another look at Evaluation Standards 4–10 and their outcomes. As you study those outcomes, remember that in this text we divided our consideration of **Mathematical Connections** into two parts: connections within mathematics and connections outside mathematics. It may also help to reread the outcomes under **Mathematical Connections** for grades 5–8 and 9–12.

Have you located all the corresponding evaluation outcomes? The first outcome under Evaluation Standard 4 and the last two under Evaluation Standard 10 seem to correspond to the "connections outside mathematics" aspect of Curriculum Standard 4. Did you identify the final sentences under Evaluation Standards 4 and 8 as corresponding to the "connections within mathematics" aspect of Curriculum Standard 4? Just as was the case with the curriculum standards, there is an overlap of outcomes in the evaluation standards. For example, there are outcomes under Evaluation Standard 4 that deal with reasoning and communication, each of which has already been treated in separate evaluation standards. There are also valuing, appreciating, and disposition outcomes under several evaluation standards although there is a separate standard devoted to disposition, Evaluation Standard 10. Part of the overlap can be explained as the result of the desirable integration of standards; part may be due to the inclinations of different groups of authors assigned to write different sections of the *Standards*.

Activities

5-1. For either the 5–8 or 9–12 grade levels, compare the outcomes of the first three curriculum standards with the outcomes of the corresponding evaluation standards. What differences exist between the sets of outcomes? Are these substantive differences? Share your conclusions with a colleague and discuss the implications of your findings.

5-2. Study the outcomes under Evaluation Standard 10: **Mathematical Disposition**. What are the relationships of these outcomes to those of the first four standards?

5-3. Study the outcomes under Evaluation Standard 4: **Mathematical Power**. How alike or different are the outcomes in this standard from those related to reasoning, communication, and problem solving?

Did you find that there wasn't really much difference between the outcomes of the first three curriculum standards and their corresponding evaluation standards? That finding makes a lot of sense. If the planning objectives are drafted in terms of terminal student outcomes, they should be the same as the evaluation objectives. For example, suppose the teacher of the prime/composite plan (pages 132-136 in the Reader) had included as one of the objectives: *The students will discover the relationship of checker piles to the factors of natural numbers*. That teacher certainly hopes this will happen, but this is not a terminal outcome. Moreover, the student can hardly be graded on whether a discovery took place. (For background on the nature of a terminal objective, see pages 108–110 in *Secondary Mathematics Instruction*; for examples of terminal objectives that reflect various levels of cognitive complexity, see page 113; for examples of objectives from, and background on, the affective domain, see pages 118–120.)

Why are there just seven evaluation standards related to student assessment when there are 13 curriculum standards at each grade range? Furthermore, none of the seven evaluation standards is labeled to correspond to curriculum standards 5–13. The key word in the preceding sentence is *labeled.* Actually, Evaluation Standards 8 and 9 encompass all the concept and procedural outcomes of those nine curriculum standards. Evaluation Standards 1–4 and 10 seem to deal with the assessment of the integration of knowledge, the ability to go beyond that knowledge, and attitudes toward mathematics. In other words, the NCTM authors are not denying the necessity to assess specific unit objectives, but they have drafted the evaluation standards in terms of a synthesis of separate unit objectives.

Activities

5-4. Study the outcomes under Evaluation Standard 8. How do they match the specific objectives for the sample plan on primes and composites found on pages 132-136 of the Reader?

5-5. Study the outcomes under Evaluation Standard 9. How do they match the implied objectives for Mrs. Armstrong's lesson on solution problems (See pages 127-128 in the Reader)?

You should have found an unusually good match of the outcomes of Evaluation Standard 8 to the objectives in the sample plan on primes and composites. Both the standard and the plan include statements having to do with verbal association learning, even though student success in achieving that type of outcome is not evidence of student understanding of the related concepts. Both refer to the student's ability to model the concept in a concrete way. In the case of the plan, the reference is to an operational definition in terms of checker piles.

"It should be evident that procedural knowledge is intertwined with conceptual knowledge" (NCTM, 1989, p. 228). This assertion, along with the outcomes under **Mathematical Procedures**, could be a paraphrase of material in *Secondary Mathematics Instruction* on rule/principle learning and planning. Did you find a similar correspondence of outcomes to the implied objectives in the outline of Mrs. Armstrong's plan?

Spiraling is an expectation for all seven student evaluation standards. Another expectation is that in any single lesson, the teacher might be assessing outcomes representative of more than one standard. Thus, the evaluation standards should be assessed in a variety of ways, over a series of lessons, and in multiple contexts. In their introduction to the section on student assessment, the NCTM authors characterized assessment as continuous, dynamic, and often informal. They reminded us that teaching and learning are interactive processes that lose their effectiveness if the teacher isn't listening to what lies behind students' responses, isn't obtaining further data on their thinking, and isn't acting responsively on the basis of these data. The authors said:

> Only through explicit and careful assessment of **how** a student does mathematics can instruction be tailored to individual needs, thereby enhancing a student's chances for success. (NCTM, 1989, p. 204)

The **Vision**, then, is definitely beyond testing.

A Bridge to the Classroom

This section begins in the real world of the classroom teacher of mathematics, in particular, in the real world of the beginning teacher. What are this teacher's first concerns, and how do evaluation issues fit into those concerns? That's easy! The very first thing one has to do before facing a class is to plan a lesson and the next is to implement the lesson in an effective manner. Therefore, the first area to be considered in this section is the role of evaluation in daily planning and instruction. What is the next concern of the beginning teacher? Very shortly after teaching a few days, some sort of assessment that leads to a grade must be designed. Thus, the second area to be considered in this section is the role of evaluation associated with grades. Finally, although pure diagnostic measures are important evaluation types, they are not a first concern of the beginning teacher. Nevertheless, sometimes a student response signals a need for diagnostic evaluation. What are some examples of simple diagnostic measures that might be used by the teacher? How can the teacher help individual students realize the intent of diagnostic measures so that these students are responsive to remediation or redirection, rather than fearful of admitting lack of understanding? These are some of the questions to be considered in the final sub-section of this chapter.

Instructional Feedback

Evaluation in daily lesson plans is known as instructional feedback. This kind of feedback is often informal, *i.e.*, not necessarily structured as a scale, form, or set of written questions. However, such feedback cannot occur only when there's extra time or if the teacher happens to think about it. As the authors of the *Standards* emphasized, this kind of evaluation vitally affects the effectiveness of teaching and learning. Study the *Instructional Model* that is reproduced here from the title page of the chapter on feedback in *Secondary Mathematics Instruction* (Figure 5.2). Notice that the feedback component is shaded, and that arrows from that component are meant to depict the possible teacher decisions made during the lesson.

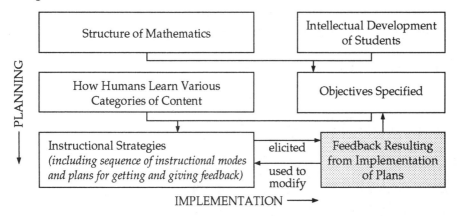

Figure 5.2: Instructional model

Now reflect on the meaning of the component labeled *Instructional Strategies*. The description under this heading includes not only instructional modes, but also feedback strategies. In other words, basic feedback strategies **must** be part of the plan.

Activities

5-6. Refer to Figure 5.3.

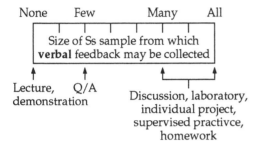

Figure 5.3: Verbal feedback potential

Explain why supervised practice and labs are among the modes characterized as having high feedback potential. Why does the question-and-answer mode have low feedback potential? What feedback potential would cooperative learning groups have? What feedback potential would in-class work of student pairs on a computer terminal have? Be sure to give a rationale for your responses.

5-7. Despite the low feedback potential of question and answer, that potential can be enhanced in several ways. The most obvious is to spread questions so that many students are involved in the process. How can the straw vote, wait time, follow-up contingency questions, or even whole-class written responses to a question be used to improve the feedback potential of question and answer? (See the suggestions on pages 8–10 and 32 of *Secondary Mathematics Instruction* for ideas.)

5-8. Study the sample supervised practice lesson on page 175 of the Reader and the sample probability laboratory lesson on pages 145-146 of the Reader.

Supervised practice, lab work, and discussion groups have high feedback potential. What feedback-getting strategies were used by the teachers in these lessons to get and keep students on track, to identify points of confusion, and to assess competence?

Multiple straw votes during a developmental question-and-answer sequence can actually interfere with the lesson. Why? As always, the teacher must think about the purpose of the developmental question-and-answer and the potential negative effect of this kind of feedback strategy. That means that there are times when it is important to collect feedback from a wide sample and other times when the collecting process might actually interfere with the teaching or learning process. When is it necessary to get feedback during a lesson, rather than the next day? Here is a starter list.

> when there are verbal (questions or answers) or nonverbal (raised eyebrows, heads turned away from demo, *etc.*) signs that confusion exists

> when the next part of the lesson depends on student understanding of the work thus far (*e.g.*, before supervised practice on the procedure just presented or before introducing a related concept)

> when students are beginning to investigate a problem via a lab, discussion groups, a work sheet, or a computer terminal—to start them on the right foot

> when students have spent some time on any of the preceding—to keep them on target

> when students are primarily observing a film, computer presentation, or multi-step demo; listening to an audio tape or a student presentation; or writing a summary of group work or their responses to a creative *What if* mathematics question

With the exception of the first situation on the list, all of the others involved planned strategies. Thus, an effective plan would have included feedback strategies responsive to each situation. For example, when students, in small groups or individually, are set to work on an investigation or even practice problems, the teacher should plan to make a quick and systematic tour of the class, listening and looking for problems with directions and organization. Next the teacher, having already analyzed the content under investigation, would begin moving from individual to individual, or group to group, looking and listening for specific content-related moves and mentally recording unexpected errors or unusual mathematical excursions. Comments, questions, and cues by the teacher during these tours are examples of giving feedback. What other planned feedback-giving strategies should occur?

Activities

5-9. Refer to the *Events of Instruction* in Table 4.1 in this text. Identify all the instances of giving feedback in the listing of the *Events of Instruction* and the rationale for each.

5-10. Some categories of feedback are praise, judgment of the validity of a response, and further explication of material. Identify examples of each of these in the probability laboratory lesson on pages 145-146 of the Reader.

Praise can be verbal or nonverbal. As teachers, it is important to develop a repertoire of such feedback-giving skills. We can say: "Good," "Well-done," or "Super job!" These are the usual comments associated with praise. However, we might also say: "Carmen's question is worth talking about. Max, how would you react?" Now Carmen has been praised and a content value has been placed on the importance of her contribution. Moreover, there's a good chance that the attention of other students will be more focused on the content and that some thinking will be generated among other students. The opportunities for praise are endless if we consider both the need to respond to a student and the possibility of building on that student's work. Think about the nonverbal signals that you recognize as a sign of approval. A smile, a "V for victory" sign, a pat on the shoulder—all are ways of communicating positive feedback. (For other examples of praise or positive reinforcement, see the heading "Praise and reward" on page 359 in the Index of *Secondary Mathematics Instruction.*)

The *Events of Instruction* provide us with a generic template for any lesson and help us to identify those points in the lesson when giving and getting feedback is essential. Suppose the objectives of a lesson included outcomes related to **Mathematics as Problem Solving, Mathematics as Communication, Mathematics as Reasoning,** or **Mathematical Connections**. Now we must be more specific about the kinds of feedback we need to get in order to assess student progress toward the outcomes. Let's see how that might work.

In the following illustrative lesson, the teacher passed out a supertangram shape sheet (Figure 5.5), and a problem sheet (Figure 5.4) to each student. The students were organized into groups, given a package of supertangrams, and shown the location of additional materials—graph paper, rulers, scissors, tracing paper— on a work table.

Problem Sheet for Supertangrams
Name _____

Read the problem. Make some notes as to possible solutions.
Consolidate ideas with other members of the group to arrive at more than one solution.

PROBLEM:

A. Develop one or more strategies to find out the relationship of the areas of the individual supertangrams to one another. With members of the group, discuss way(s) to demonstrate the validity of each strategy.

Record individual strategies and results here.

(More space would be left.)

If your group is satisfied that they have responded to A as fully as possible, then read B and talk about the solution process and the solution.

B. Order the supertangrams by perimeter, smallest to largest. Use the numbers assigned to the shapes on the shape sheet. Talk about ways to defend (prove?) the ordering process.

Record individual strategies and results here.

(More space would be left.)

Figure 5.4: Supertangrams problem sheet

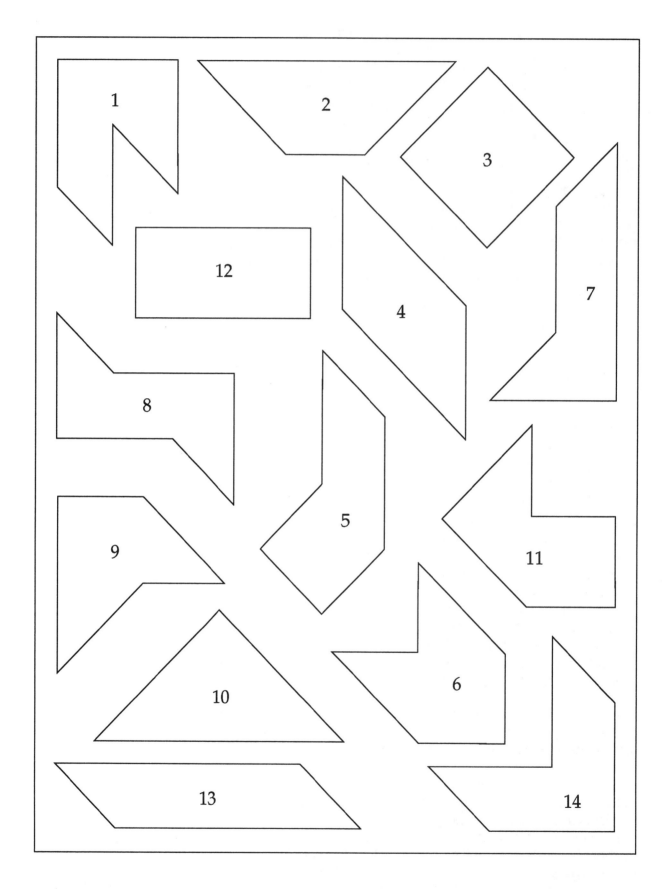

Figure 5.5: Supertangrams shape sheet

Activities

5-11. With a colleague and appropriate materials, work out part A of the supertangram lesson.

a. What are the specific problem-solving outcomes that are reasonable to expect of the students? Describe the possible feedback (both positive and negative) the teacher might collect while touring systematically and observing.

b. What outcomes related to **Mathematics as Communication** and **Mathematics as Reasoning** would be associated with successful implementation of this lesson? How can the teacher get feedback during the lesson on the extent of student progress toward these outcomes?

c. To what extent are **Mathematical Connections** being addressed in this lesson?

Here are the objectives listed by a teacher who taught the supertangram lesson. Compare these with your responses to Activity 5-11.

1. Correctly apply one or more strategies (*e.g.*, folding and/or cutting paper copies to get a match) to the area problem.

2. Give justification for results in terms of deductive reasoning.

3. Identify and use the mathematical relationships of each strategy (*e.g.*, folding strategy and reflections and thus, congruence and equal areas).

4. Explain to other students the mathematical basis of an approach by using words and by demonstrating with materials.

This teacher wasn't sure the students would have time to work on the perimeter question so no specific objectives were listed for this part of the lesson. All four of the generic standards are reflected in this set of objectives. Those related to communication (objective 4) and reasoning (objective 2) are probably the easiest to identify. The other two objectives (1 and 3) certainly fit in with the problem-solving outcomes of Curriculum Standard 1, but they also correspond to the first two outcomes in the grades 9–12 set of Curriculum Standard 4. Check those Curriculum Standard 4 outcomes now and compare them with objectives 1 and 3 from the supertangram lesson. Why are connections outcomes related to these two objectives? The key is the concept of representation, which is used in both of the designated outcomes from Curriculum Standard 4.

What kind of instructional feedback would be needed to ascertain progress toward these objectives? If you were the teacher, you would be listening to arguments and ideas, observing the techniques students used to gather data or make a point. Who chose to trace the supertangrams on graph paper and meticulously calculate the number of squares within the regions? Did others trace the supertangrams on tracing paper, cut them out, fold and match sections? What conjectures did they make and why? Because students in the grades 9–12 sometimes believe that using concrete materials is seen as undesirable, you might have to give feedback repeatedly on the availability of the materials and the desirability of more than one strategy. A way of encouraging those using tracing paper is to quietly ask them what they are trying to show and to be positive about the potential of the approach. It would not be a good idea to end their explorations by judging the methods they are using. You should, however, be concerned if any of the students were making an argument based on concrete measurements alone, or on the apparent correspondence of two folded shapes according to a visual test. You should also be concerned about the misuse of words such as *prove*. If any of these errors should occur and are not corrected by other members of the group, you would need to ask a question or provide a reminder. Your role would be to prod the students to identify the weakness in reasoning and the need to go beyond special cases in proof.

Depending on the students' mathematical background, you'd also expect to hear language such as *reflection, mirror image, rotation,* and *translation* when students are describing concrete movements of one figure or section that led to a correspondence with another figure or section. If instead they are using language such as *flipping, folding, turning,* and *sliding,* their productive thinking should not be interrupted to get them to use precise mathematical vocabulary. That's feedback that may be withheld until the summing-up part of the lesson when all groups are reporting their findings. At that time, as student groups share their final reports, they should be expected to react to one another's ideas, reasoning, and language. Lapses in vocabulary may range all the way from wrong language that reflects erroneous concepts (*e.g.*, using *prove* when a conclusion is reached by inductive reasoning) to formal versus intuitive language (*e.g.*, reflecting instead of flipping). The former is serious and needs to be attended to at the time you hear it being used. The latter should not even be approached as an error, but rather as a gradual movement toward more precision in terminology.

In what sense does the problem in the supertangram lesson match the criteria for a novel problem? Notice that the design of the lesson gives the students an opportunity to share strategies, to try out heuristics, to reflect on approaches. If the students have not had much experience in lessons of this type, the teacher may need to bolster confidence, suggest use of concrete materials, and ask leading questions. If, however, past lessons have been devoted to problem-solving approaches and problem-solving attitudes, the teacher might wish to use an observation record form to assess student progress in those areas (Figure 5.6). During a single lesson, the teacher could rate a sample of students by using marks, such as $\sqrt{+}$, $\sqrt{}$, and $\sqrt{-}$. The results of this form should not be used for grading in the early stages of the teaching of problem solving. However, problem-solving outcomes eventually should become part of the total grade. Then, these results may be used for grading purposes. How

Behavior	Lisa	Sam	Meghan	Carlos
1. Suggested 1 or more approaches				
2. Used concrete materials to get data on own initiative				
3. Made 1 or more conjectures				
4. Questioned rationale of others				
5. Moved the shapes in systematic ways				
6. Began making a systematic record				
7. Argued from mathematics				

Figure 5.6: Sample observation record form

is this managed? We'll return to this question in the next section on evaluation for grading purposes.

In the supertangram lesson, the students need to be able to build on some of the connections that exist within mathematics. In the sunflower seed lab summarized on page 165 of the Reader, there are obvious references to connections between mathematics and science and between mathematics and the everyday world of the students. In this lesson, middle grade students working in pairs count the black stripes on both sides of a sunflower seed. Seeds are placed in one of 14 cups, numbered 0 to 13. Pairs of students can perform this activity on about 200 seeds rather quickly. When they are finished, the cups are emptied into 14 identically numbered hygrometer jars on the front desk. As the students add to the hygrometer jars, a three-dimensional representation of the normal curve appears. In the follow-up question-and-answer and mini-lecture period, the concepts of frequency, range, median, mode, normal curve of distribution, variation in nature, sampling, and mathematical model can be explored. The teacher's objectives for this lesson might include the following.

> Explain the reason why data from living things might not form an exactly symmetrical curve even though the characteristic is said to be normally distributed.

> Identify other characteristics of living things that are normally distributed.

> Given a collection of data, be able to identify the median and mode.

> Write a short paragraph about the meaning of *sampling* and the effect of the kind of sample on results.

The instructional feedback strategies used by the teacher in this lesson would not differ greatly from those used by the teacher in the supertangram lesson. However, the nature of the feedback for which each was listening and looking would be different. There is nothing to prove in this lesson. It would be important, instead, to ensure that students were observing spatial results at various stages in the process. After counting almost half of the 2 seeds, what seems to be the relationship of the amounts in the cups? Is this conjecture supported as more seeds are counted? As each pair of students adds the contents of their 14 cups to the hygrometer jars, what seems to be the relationship of the amounts in the jars? Are these conjectures supported as more student pairs empty cups? The teacher will have to cue, direct students' attention to the spatial differences, ask that conjectures be made, and ask that the early conjectures be tested. (If the teacher has more than one class doing this lab, further seeds can be added to the jars later in the day.) The language of *samples* and *probable* will be used by the teacher. In the summary part of the lesson, the teacher will assist the students to use ideas from the everyday world and science class when asked for other examples of traits that are normally distributed. A out-of-class assignment could be to find pictures of other examples of natural variation in humans, plants, or other animals.

A great many lessons will be focused on the learning of concepts or rules. Sometimes these same lessons will also be focused on the learning of specific psychomotor skills or, more often, the learning of verbal associations. In the earlier part of this chapter when considering Evaluation Standards 8 and 9, you were referred to previous work on concept and rule/principle learning and planning. Then you were reminded that the test for ensuring that a concept has been learned is student ability to generalize the concept to a variety of specific instances not previously used, while the test for ensuring that a rule/principle has been learned is student ability to demonstrate instances of the rule in a variety of situations. Statements, definitions, and labels associated with concepts and rules are examples of verbal association learning. The test for that kind of learning includes :

> give name and elicit statement of idea or description of object; have students use in context (Farrell and Farmer, 1988, page 167).

What kind of instructional feedback should the teacher be collecting in lessons with these kinds of mathematical outcomes? First, it is important to **specify** the outcomes, as was done in the prime/composite lesson on pages 132-136 of the Reader. The benefit in making the outcomes crystal clear is three-fold. First, it's easier to design instructional strategies that will help students reach those learnings. Second, it's easier to design follow-up evaluation that will be fair to the students. Finally, knowing the terminal objectives and the instructional strategies, the teacher can reflect on the kind of feedback that signifies progress at various points of the lesson. For example, when making a tour during the checkers lab, the teacher will be looking for specific arrays of checkers and scanning lab sheets for key responses. Deviations will be noticed quickly and the teacher can stop to ask a question or set a student on the right track.

Study the entire prime/composite plan with particular attention to the feedback columns. Notice how feedback on specific verbal associations (labels, a formal definition) is delayed so that students will not erroneously confuse the concept with the label. When the students are asked to attach labels to the checker numbers, the teacher works on both verbal association learning and concept learning. How? Why does the teacher choose such large numbers in the *generalizing to other cases* section of the plan? How does the teacher get feedback that most or all students can generalize to other cases? Did you notice the shift from the question-and-answer in the previous part of the plan to a strategy in which each student writes answers and reasons? In each case, the feedback strategies are directed toward getting data on the extent of learning of as many students as possible. (For further details on these strategies, refer to pages 165–167 of *Secondary Mathematics Instruction*.)

Instructional feedback affects both the planning and implementation phases of the Instructional Model (Figure 5.2). However, the model depicts only the actual feedback received by implementing the feedback strategies. In what sense is feedback considered in the planning of the lesson? The teacher reflects on the kind of feedback desired at each stage of the lesson. Thus, in a tour during homework post-mortem, the teacher wouldn't just look at papers for evidence of some work, but at one or two specific examples for evidence of the use of a common misconception. As in every other chapter of this text, teacher reflection on what is to be taught and what is to be learned is a critical ingredient in effective instruction.

Evaluation Associated with Grades

A first principle is that no objective should be assessed for a grade if feedback collection during the lesson indicates that the lesson was not effective. If this were the case, instruction and/or objectives must be revised (see Figure 5.2). Notice how the concept of *effectiveness of instruction* depends on the quality of the interaction between teacher and students, and that quality is monitored by well-designed feedback strategies. We'll assume that both the supertangram lesson and the sunflower seed lesson were judged to be effective. What follows are items the teachers designed to assess achievement of one or more of the lesson objectives for each lesson.

Supertangram Lesson Items:

1. Use the regular hexagonal cut-out (supplied with item) and consider it one member of a set of H-grams. Create (draw) new members of the set, all having the same area as the hexagon. Label each new member and write down the strategy you used to create it. (You may use any materials from the work table in working on your response.)

2. Given two irregular polygonal shapes, how would you prove or disprove the assertion that they have the same area?

3. Design a plan to create a set of polygons, all of which have the same perimeter. (Materials from the work table may be used to help you design your plan.) Investigate the areas of the set you created.

Sunflower Seed Lesson Assessment:

Have students form a living graph of their heights by standing in a row, from smallest to tallest. Equally tall students should stand one in front of the other. The teacher should model the situation on the overhead or chalkboard, with names of students listed next to the appropriate picture on the mode. Have the students write responses to the following questions :

1. Describe or name the median for the graph of this class by heights. Explain why you made that selection.

2. Describe or name the mode for the graph of this class by heights. Explain why you made that selection.

3. Describe or name the range for the graph of this class. Explain why you made that selection.

4. The graph of this class probably doesn't look much like a normal curve. Would we get a graph that looks more like the normal curve if we stood all the eighth graders in the school by height? Why or why not?

Activities

5-12. For the supertangram lesson, which assessment items would provide evidence of student ability with respect to one or more of the teacher's objectives for that lesson? Justify your answer.

5-13. Which question(s) from the supertangram assessment would provide evidence of student ability with respect to **Problem Solving, Communication, or Reasoning**? Be specific about the evidence and the outcomes related to that evidence.

5-14. How does the assessment from the sunflower seed lesson match the teacher's objectives? How are particular outcomes from **Mathematical Connections** assessed by the question(s)?

5-15. Choose an assessment item for each lesson. With a colleague, prepare a model (or keyed) response for each. Then decide how to apportion credit for a student's response. (Refer to pages 217–219 in *Secondary Mathematics Instruction* for suggestions.)

The items in both sets require some kind of written or oral response. Oral response? Oh, yes! Don't discount the possibility of a pair of students working together on a question and collaborating on an oral response to the class. Students should be encouraged to supplement the oral report with overhead acetate drawings, board outlines, or models, when appropriate. They should be prepared to respond to questions from the class. At the same time, the class has to be prepared to listen responsively, perhaps to take notes, and to question areas of confusion or interest. They will be convinced of the importance of their role when the teacher uses ideas presented in the oral reports in homework assignments, introductions to related topics, and perhaps in later evaluation measures. Remember evaluation is a **part** of instruction; it is **not** separate from it.

Oral reports, reports of library research work, and reports on the design of a model and the model itself might be assessed on a scale that has already been shared with students (Figure 5.7).

Oral Report

Name_____   Partner_____
Question or Topic_____   Date of Report_____

Criteria	Scale				
1. Quality of organization	1	2	3	4	5
2. Clarity of communication	1	2	3	4	5
3. Validity of mathematical ideas	1	2	3	4	5
4. Demonstration of mathematical connections	1	2	3	4	5
5. Thoroughness of study	1	2	3	4	5

COMMENTS:

Figure 5.7: Sample scale for oral reports

The meaning of the criteria can be explained prior to the presentations, by reference to a different class activity. Some teachers like to leave room for comments after each of the criteria. For example, under the fourth scale item, the teacher might write: *You made a nice use of symmetry when explaining the folding shortcut.* The teacher should indicate whether a rating of 5 is high or low and needs to communicate to students the way in which these scale ratings will be part of a summative grade for the unit, quarter, semester, or year. (For background information on this issue, refer to pages 238–239 of *Secondary Mathematics Instruction*. Also see Figure 8.8 on page 237 for an example of a summative observation record form to assess affective objectives during a unit on similarity.)

Could **Mathematical Power** be assessed by means of a scale? Yes, it could. However, assessments of student progress toward the standard of **Mathematical Power** occur over time and in conjunction with the evaluation of other standards. Reread example 1 from the supertangram assessment set. Suppose a student's response showed an analysis of supertangram relationships and included a rationale for extending some of those relationships to create a set of H-grams. That student is demonstrating an aspect of both problem solving and mathematical power. This illustration is an indication of the need to collect data in the form of anecdotal records, scale ratings, student project results, and responses to higher level cognitive items in an individual student folder.

In Chapter 2 of this text, several activities included ways to develop listening, reading, and comprehension skills. In Activity 2-8, students listened to tape recordings and, in one case, had to complete a diagram from the instruction. In the second tape recording, they were to analyze a student's explanation of a mathematical idea. If the feedback indicated that these developmental lessons had been effective, similar exercises could have been developed for the purpose of grading the students in the area of **Mathematical Communication**. For example, two transcripts for giving directions to an absent student might be given to the students. They should be directed to read and evaluate both by responding to a set of questions, such as the following.

> Does either transcript leave out anything important or include misinformation? Give evidence.
>
> Is the mathematical language correct and appropriate? If not, identify the problem words, phrases, or symbols.
>
> What would you like to change if you were giving directions for this assignment over the phone? Why?

Activities

5-16. Assume you are going to rate the students' responses to the transcript analysis item.

a. What evidence would you use to judge whether a student has read with comprehension?

b. What evidence would you use to judge the extent to which a student has expressed ideas clearly and accurately?

5-17. Read the geoboard exercise described here. Then respond to questions **a**, **b**, and **c**.

A 25-peg geoboard is handed to each student. Each student is also given 2 rubber bands, a test sheet, and a sheet of dot paper. Students are told to read the information and follow the directions on the test sheet.

From the Test Sheet

Geoboard Item: Use a rubber band to place a square with sides of 4 units on the board. Then draw the square on the dot paper. Start at the lower left corner of the square on dot paper and label that dot "1". Continue up that side and number the dots in order "2", "3", "4", and "5". The dot in the upper left corner should be numbered "5". Continue around the square, numbering the rest of the dots. (Note: the upper right corner should be numbered "9". The last dot is numbered "16".) Now use the second rubber band on the geoboard to form a second quadrilateral by positioning the corners at pegs corresponding to dots numbered "2", "6", "10", and "14". Copy this figure on the dot paper at the correct locations.

1. What is the area of the first quadrilateral? _____

2. After constructing the second quadrilateral, how many shapes are inside the first quadrilateral? _____

3. Four of these shapes are congruent. Explain why.

_____

4. Find the area of the second quadrilateral. _____

5. What else can you find out about the second quadrilateral?

_____

a. What evidence will the teacher be able to use to judge whether the student has read with comprehension? whether the student is able to check on his or her own reading comprehension in an intelligent way?

b. Which questions will provide evidence on the extent to which the student can express ideas clearly and accurately?

c. What standards, other than communication, can be assessed from responses to this item?

In Activity 5-17, the students must follow written directions in order to correctly place appropriate figures on a geoboard. The teacher can scan geoboards while administering the assessment measure. However, since this is not supervised practice but evaluation, it would be inappropriate to tour systematically. Not only might that kind of behavior lead the students to assume they can ask for help, but it puts the teacher in a poor position for administration of a measure to be graded. (For more information on test administration, refer to page 324 of *Secondary Mathematics Instruction*.) When the papers are handed in, the teacher will be able to examine the figures on the dot paper and the responses on the work sheet. Did you notice that the students were given some information in question 3—information that might have helped them to check and change the geoboard figures as needed. What other helpful information was given them?

Thus far, you've been given some examples of different kinds of evaluation measures for both the cognitive and affective domains. An example of a rating scale has been presented. You've been given examples of different ways of assessing the same outcomes, as well as how a single measure might be used to assess more than one major outcome. On a daily basis, it's relatively easy to make sure that assessment measures correspond to the outcomes of Evaluation Standard 1: **Alignment.** For example, two of the supertangram lesson assessment items were designed to allow students to use concrete materials, as they had been able to do in the supertangram lesson. The sunflower seed assessment was based on a living graph, which was then modeled by a pictorial graph on the board. This concrete and active mode corresponded to the lab activity of the lesson itself. However, a more systematic consideration of **Alignment** is needed when the teacher has to design a unit test. We recommend that the teacher use a table of specifications.

Activities

5-18. Read about constructing and using a table of specifications on page 176 of the Reader.

a. Why didn't the teacher list the actual objectives from the individual plans, rather than the major topics of the unit?

b. There are at least two ways in which the first four curriculum standards might be included in this table of specifications. With a colleague, talk about how to incorporate the first four curriculum standards.

The authors of the *Standards* give examples of many of the grades 9–12 outcomes at different levels of complexity. Similarly, the table of specifications referred to in Activity 5-18 shows a breakdown of objectives according to Bloom's Taxonomy, even though the specific objectives are not stated in the table. No matter which classification system is used, the teacher needs to be sure that evaluation items are not weighted unfairly or inappropriately. Did you consider adding the first four curriculum standards to the top margin of the table? If you did, you'd test these in the context of one or more of the content topics. Suppose the standards were added to the column of content topics. What would this mean? The issue of alignment must be considered each time the teacher assesses a unit, and again when it's time to decide on a composite grade at the end of a quarter, term, or year.

Diagnostic Evaluation

The word *diagnosis* has already been used in earlier chapters of this text. In Chapter 2, writing tasks were characterized as valuable diagnostic tools. In Chapter 4, ongoing diagnosis of possible student misconceptions was considered essential if mathematical connections were to be learned. In these cases, the distinctions between instructional feedback and diagnostic evaluation were often blurred. One way of distinguishing between these two types of evaluation is to recognize that instructional feedback is focused primarily on the class. Data from the class are sampled by feedback-getting strategies, and on-the-spot instructional decisions are based on an analysis of what is best for the class. Diagnostic evaluation is focused primarily on the individual. Diagnostic evaluation goes beyond the identification of student errors to the identification of the causes of those errors. Thus, during a supervised practice tour, the teacher may become aware that one or two students seem to be making consistent, rather than random, errors. Diagnostic evaluation will be needed in order to identify the source of the errors. Some ideas for this kind of evaluation have been provided in Chapters 2 and 4 of this text. These include probing questions, journal entries, concept maps, and essays. (For others, see pages 39–40, 208–209 and 236 of *Secondary Mathematics Instruction*.) To be involved in diagnostic evaluation is to pay attention to individual differences. Individual learning styles, different levels of cognitive development, preferred ways of processing information, diverse backgrounds—all of these differences affect learning in one or more ways.

As noted earlier, the very attention being paid to an individual in trouble can sometimes backfire. The teacher has to be careful to treat the diagnostic activity in a positive way. It is too easy for the student to feel stigmatized by extra attention on a learning weakness. A student's incorrect thinking can often lead to greater learning on the part of the rest of the class if all cooperatively work on the miscues—from the student perspective—that led to the misconception. A systematic incorrect procedure is not due to carelessness, but to a reasoned approach. The teacher needs to be aware of the affective aspect of diagnosis and communicate the view that honest errors are not a sign of a dumb student.

To be involved in diagnostic evaluation is also to reflect on the structure of the mathematics being taught. How does a student think of that concept? What does a student "hear" when a teacher speaks in mathematical language? How does a student's cognitive development affect learning of proportions, spatial reasoning, ability to engage in deductive reasoning? These questions have been referred to earlier in this text. On a less sophisticated level, we would not expect a lesson in calculus to be effective for a student whose learning of functions was deficient. It doesn't really matter that the student was in a course in which the topic was "covered." What matters is whether the student learned the necessary concepts. From a cognitive development point of view, we would say that adaptation can't occur under those circumstances. (Refer to pages 59–60 of *Secondary Mathematics Instruction* for an explanation of the functioning of cognitive interaction, the need for diagnosis of prerequisite learnings, and some other examples of informal diagnostic measures.)

The evaluation section of the *Standards* often seems overwhelming to the individual teacher. However, we should view it as one more essential part of the vision, which must begin with individual teachers, be supported by supervisors and administrators, and be fully realized by attention to the program evaluation standards. For our part, it is important to begin.

References

National Council of Teachers of Mathematics. 1989. *Curriculum and evaluation standards for school mathematics.* Reston, VA: NCTM.

Farrell, M.A., and Farmer, W. A. 1988. *Secondary mathematics instruction: An integrated approach.* Dedham, MA: Janson Publications.

Chapter Six

A Look Ahead

You've done your part. You've studied sample lessons in the Reader, and in the text. You've examined theories and research in background readings, as they related to a particular standard. You and one or more colleagues have talked about ways to design a lesson or an evaluation measure. If you've been able to combine teaching experience and these reflections, you may have tried out some of the suggested strategies. If you have not yet taught, you will want to return to this text as you plan lessons. Remember, thoughtful reflection of this kind is a prerequisite for, not a guarantee of, survival as an effective mathematics teacher. It's up to you to implement the ideas and to continue the process of improving your own teaching.

What are some of the concerns that you will face as you try to implement the vision of the *Standards*? The most immediate concerns will be integration, articulation, communication, evaluation, technology, and differentiation. What do each of these mean and how can you deal with each?

Integration

The NCTM authors made it quite clear that the first four curriculum standards should not be isolated for treatment, but should infuse the content standards at each grade range. However, the manner in which the integration occurs can actually minimize the impact of those four major standards. What a paradoxical notion! How can that happen? Just return to a sample lesson that has been the subject of several questions in this text—the probability lab from pages 145-146 of the Reader. There is no doubt that this lesson has the potential for achieving communication objectives, as well as objectives in the content area of probability. However, if the teacher never articulates the communication objectives, it is unlikely that feedback strategies will be designed to assess progress toward those objectives. The incidental communication that occurs will not be assessed or redirected. The potential will be lost.

Therefore, you have to integrate, but not submerge, the generic standards. This kind of integration is always an instructional balancing act. A first necessary step is to specify **all** the objectives of the plan. Over time, it will be important to talk with your cooperating teacher or department head about the relative emphasis being placed on generic and content objectives. Other teachers, who may not be as familiar with the *Standards* as you are, may not realize that some of their favorite lessons may be excellent vehicles for this kind of integration. Here's where you can help. Be alert to the potential of lab lessons or small-group work sessions for instruction in any of the first four standards. When the opportunity arises, be ready to share ways of stating and assessing the first four standards.

Articulation

From the curricular point of view, it's assumed that the objectives for the middle school youngster are built on those to be attained by the primary school child. Similarly, the grades 9–12 standards have been developed to follow the grades 5–8 curriculum. Suppose your sixth grade students were not in a primary school environment attuned to the *Standards* or that your tenth grade students had no previous experience at any level in standards-type classrooms. Both of these situations are likely to be realistic occurrences for the near future. It takes time for reforms to work their way into classrooms. Does this mean that you must abandon attempts to design lessons focused on the first four standards? Definitely not! However, you'll need to compensate for the novelty of lessons emphasizing communication, problem solving, reasoning, and connections. Students whose previous mathematics classroom experience has been solely to listen may resist a more active role. Some students are threatened by requests to keep a written log of their feelings about lab or project work. Even though such reports might have been used in science class, the behaviors do not easily transfer to mathematics class. In their experience, the rewards in mathematics class were for exact recitation of text definitions and rules. The students have to be convinced that the ground rules have changed and why.

If you are aware of the previous experience of the students and the potential for attitudinal problems, you can guide students into their new roles. That means it's important to consult with other teachers about the previous classroom experience of the students. If you are student teaching, you should observe the mathematics class to which you've been assigned as a first teaching experience. Then talk with the cooperating teacher about how you might introduce small-group work or labs or any other student-centered mode. You need to be aware that veteran teachers may feel threatened by the more student-oriented modes. However, they may be quite willing to help a student teacher work through a lab class so that they can learn from the efforts and ideas of a person new to the profession.

There are other important articulation issues, and many are directly related to content objectives in the *Standards*. You will need to study the related standards at the various grade levels and adjust your plans for any missed material. That will not be easy. You will also have to adjust your long-term goals. Certainly, if your tenth graders did not have instruction geared toward a specific set of content standards in the primary and middle grades, you will have to help them master these prerequisites before you introduce related content from the standards for grades 9–12. There will have to be continual reflecting and adjusting on your part.

Communication

In the previous section, the need to communicate with students was addressed. You were reminded that the teacher needs to let students know that something new is being expected of them. The following questions must be addressed on a very concrete level, in the context of a specific lesson.

What are the outcomes that the students are expected to achieve?

Why are they considered important?

How will lessons incorporate the outcomes?

How will students be assessed on these outcomes?

Many of the answers will fall in the category of giving the students a rationale for the outcomes, the strategies, and the feedback processes. However, the rationale should not be in terms of what some national group thinks students should learn, but in terms of the relation of these outcomes to the nature of mathematics and the long-term needs of students.

Communication must not end with the students. For the student teacher or the beginning teacher, there will be the need to communicate with other mathematics teachers in the school. This kind of communication often takes place informally in the faculty room or across the lunch table. Be forewarned! You're not being asked to be a crusader among the faculty. When your presence in the school depends on the school's cooperation with the college or university in your region, it is seldom desirable to assume that you are there to educate the certified teachers. Instead, as you are asked about your progress, your classroom experiences, or the recent visit of your college supervisor, there will be opportunities to clarify some of the approaches you are using with a class, as well as your problems and successes. It's vital that you have a clear understanding of the relationship of the *Standards* to the lessons you design. Then you'll be able to explain this to others.

If you are present during a parent-teacher conference or a school open house, you should be able to respond to parental questions on the use of logs in mathematics class, the quizzes based on tape-recorded directions, or the projects in which students had to scan the newspapers and journals for math-related material. For most parents, mathematics was a subject in which a lot of memorizing was needed. They may, in fact, believe that it is important to be successful in the memory work and worry that their children are being short-changed. Once again, it is important that you are convinced of the educational and mathematical validity of *Standards*-related lessons and that you can explain your convictions to others.

Evaluation

The fifth chapter of this text was on the subject of the evaluation standards. What else is there to say? Remember the first three evaluation standards? They dealt with the correspondence of evaluation to the outcomes and to the instructional modes used to teach those outcomes. Although colleagues in the schools may be understanding if you use manipulatives in an eleventh grade mathematics class, some will not be at all understanding if students are provided manipulatives in a test setting. Some teachers may think you are spoon-feeding these abler students, when in fact, if you've been creative, you were able to assess more complex understandings by using manipulatives. You can put your colleagues' fears to rest by sharing both the activities and the related questions students had to answer.

Reporting of grades may pose another kind of problem. Suppose your rating of student problem-solving skills was collected on a rating scale, or that your assessment of a student project illustrating a mathematical model was in terms of comments and a final letter grade. How can these grades be converted and assimilated into the numerical percent grade used in the district? If you are a student teacher, grades have to be merged with those previously given by the cooperating teacher. What can you do? There is no easy answer. You, the college supervisor, and the cooperating teacher may need to sit down together and discuss the outcomes of your lessons, the correspondence of evaluation to outcomes, and the best way to communicate grades. The results of that discussion may well be a compromise, rather than a significant change in the way grades are reported. That is the real world of the classroom. If you have your own class, you have more independence, but you still must work around the school grading policies. If those policies are antithetical to the evaluation standards, you need to begin to lobby to get them changed. That may not be as impossible as it sounds. All major administrative and supervisory associations have come out in support of the *Standards*, so you will have a leg up on needed changes.

Technology

One of the assumptions of the authors of the *Standards* is that technology has changed the "nature of the problems important to mathematics and the methods mathematicians use to investigate them" (NCTM, 1989, p.8). For this reason, the authors emphasize the importance of widening the access of all students to calculators and computers. However, the authors point out that access alone is not enough. Teachers must design lessons and assessments incorporating the use of such technological aids. Unfortunately, calculators and computers are still not available in some classrooms. If you are student teaching in such an environment, you might be able to negotiate a loan from the area college campus for the purpose of instruction. Even if only one computer can be loaned for demonstration purposes, you can still plan parts of lessons around its use. If your lessons are well-designed, students, parents, and other teachers will learn that computers can be used in meaningful instructional ways, not just as a tool for drilling students. Software, such as *The Geometer's Sketchpad*, could be used to support student visualization, to look for patterns, or to get data to be used to guide a proof.

As with some of the other issues identified in this chapter, mistaken concepts and negative attitudes may be reasons for lack of availability of technological equipment in mathematics classrooms. Parents and school board members may believe that calculators will undermine learning by serving as a crutch to memory. They are correct if the only instructional use is to serve as a check on arithmetic processes. On the other hand, the use of calculators to explore numerical patterns or to investigate graphs of a function is a powerful argument for their inclusion as a piece of instructional equipment.

Finally, don't limit your view of technology to calculators and computers. Use the ever-present TV and its companion, the video recorder, to record programs with connections to mathematics, for later use in class. Incorporate ideas from some of the excellent instructional materials on video tape, such as "The Challenge of the Unknown." (See some additional ideas for using an array of audio, visual, and technological aids on pages 304–306 of *Secondary Mathematics Instruction*.)

Differentiation

What is meant by this label? It's really shorthand for differentiation of instruction based on individual differences. In Chapter 5 of this text, there were references to some of those differences. Let's re-examine the first four curriculum standards in terms of individual differences.

In *problem-solving* sessions, students need to be able to analyze problem components, to adapt known strategies to new situations, and sometimes to generate their own problems from a set of data. How might individual differences among students affect their attainment of problem-solving outcomes? Perhaps the work on cognitive style can give us a clue. Some students are primarily field independent. They may divide problem components up into a number of aspects and try to find a rule or tactic to match any one of these. Others are primarily field dependent and may view a problem as a whole and try to find a strategy that fits that whole picture. In the latter case, the students may be less flexible about transforming a part of the problem. Processing differences might also affect problem-solving skills. Some students are comfortable processing information given in visual, concrete, or verbal representations. They are able to represent a problem in different mental ways. Other students are more limited in the ways they process information. As a teacher, you'll observe different working styles and become aware of different ways of processing information. Sharing approaches, stumbling blocks, and the ways students interpret a situation are ways to help students build on their differing approaches.

Communication outcomes include those that seem easily acquired by many students as well as those difficult for almost all. Reading, writing, listening, and speaking are the general communication skills. Added to their usual levels of difficulty is the requirement that students must read, write, listen, and speak about mathematics. Furthermore, they must do this in informal ways, not by simply reciting or writing a memorized statement, as in the past. Even the student who is verbally superior in English class may stumble in mathematics class. The stumbles are likely to end when that student realizes that the teacher really is interested in the individual ways of expressing a mathematical process. Students who are less able with the spoken word will need to be identified and assisted in meeting these communication outcomes. A similar case can be made for each of the other communication skills. These individual differences may be due, primarily, to the variety of previous in- and out-of-school experiences of the students. How rich or poor has their communication life been? Are you dealing with students whose everyday language is not standard English, but street English, or even another language? Such students may be excellent communicators in their everyday language but may be at a disadvantage when forced to used standard English in a classroom.

Reasoning is so much a part of mathematics that individual differences in ability to reason may seem to be a matter of despair. Here is a case where differences in cognitive development will show up dramatically. We know that the same age group may include students with several levels of cognitive development, or students with the same potential level but different prior concrete experiences. (See the nested stages model on page 55 in *Secondary Mathematics Instruction*.) Lessons with multiple contexts—concrete, visual, and symbolic—help to meet the needs of students at various points in their cognitive development. Moreover, a carefully designed lesson with hands-on activity may help students at different levels of cognitive development to learn differentially. Reflect on the supertangram lesson outlined in Chapter 5 of this text. How could that lesson lead to just these kinds of results?

Finally, we turn our attention to the standard dealing with *connections*. What kind of individual differences affect attainment of these outcomes? The within-mathematics connections depend on reasoning, and thus on cognitive development, as well as on the way students process and retain information. The outside-mathematics connections are dependent on the variety of real-life experiences of the students and the teacher's knowledge and use of these in mathematics class. In an inner-city classroom, the greatest gap in individual differences is likely to be the one between the teacher and the students. In this case, the teacher must learn from the students so that the connections are real to them. However, these connections are also dependent on an understanding of the concept of mathematical model. The understanding of that very powerful concept requires a grasp of the difference between the possible and the real. As you might surmise, a sophisticated understanding of the concept of mathematical model requires formal operational reasoning. Thus, individual differences in cognitive development must be taken into account as you design lessons whose objectives include mathematical-connections outcomes.

So You Want to be a Teacher

This title was a phrase used in the section of *Secondary Mathematics Instruction* entitled "To The Student." In that section, Farrell and Farmer (1988) described the various characteristics a beginning teacher would need to bring to the profession. Among them were endurance, comprehension of subject matter, a genuine enjoyment of young people, and a knowledge of self. The reader was advised that he or she needed to be willing to learn, to interact with ideas and people, and to have the courage to put those ideas into action in the classroom. Farrell and Farmer's contribution to the prospective mathematics teacher's education was to provide an integrated approach to instruction, a model of instruction. The model served as a template for study—a way to reflect on the components of instruction in mathematics. That same model has been the basis for making sense of the goals in the *Standards.* In this text, classroom examples were merged with related theory. You were continually faced with the characterization of teaching as a problem-solving task.

If you have not yet taught, you are encouraged to return to this text after you have had some classroom experience. Former students tell us that they find new insights in the familiar material found in *Secondary Mathematics Instruction* and that the reflection helps them to repair a lesson or rethink an approach to a student. It is my hope that this text will serve a similar, long-term purpose.

References

Farrell, M. A., and Farmer, W. A. 1988. *Secondary mathematics instruction: An integrated approach.* Dedham, MA: Janson Publications.

National Council of Teachers of Mathematics. 1989. *Curriculum and evaluation standards for school mathematics.* Reston, VA: NCTM.

The Challenge of the Unknown Film Series from Phillips Petroleum Company, Bartlesville, OK.

Jackie, N. *The Geometer's Sketchpad* ver. 2.0. 1992. Berkeley, CA: Key Curriculum Press.

Implementing the NCTM *Standards*

THE READER

Selection 1

Mrs. Armstrong Lesson on Solution Type Problems

Some type problems cannot be represented by a common formula. The alcohol evaporation problem given earlier is a good example of such a situation problem. In order to solve such a problem, the student must be able to apply translation skills at two levels of complexity. In the following paragraphs, we outline the idea concocted by Mrs. Armstrong, who was beginning to plan a lesson on an introduction to solution problems. Mrs. Armstrong kept in mind the two characteristics of situation type problems pointed out earlier.

1. The situation must be meaningful *to the student* and

2. The student must be alerted to the need for an initial strategy other than translation.

The first characteristic is the key to what should occur in the construction of the advance organizer and the follow-up development of this kind of rule. The situation must be made meaningful in a concrete way. The plan itself is not included, but you should be able to expand the development described below into an appropriate sequence of instructional and feedback strategies.

"I'll start with a demonstration that can be matched to a specific problem. I'd better start with a situation in which water is added; evaporating may befuddle their thinking," Mrs. Armstrong mused.

After borrowing two 100 cc graduated beakers, two 10 cc graduated beakers, and a bottle of fluorescein dye from the chemistry teacher, she added drops of the dye to the 90 cc of water in one of the beakers until an intense color was present. She recorded the number of drops used and then observed the effect of adding 10 cc of water on the color. There was a perceptible change.

Mrs. Armstrong decided to do the final step of the demo in exactly the same way at the beginning of class and then question the students as to their observations. As a result of a last-minute brainstorm she decided to use all four beakers with the two 100 cc kind already containing identical amounts and intensity of the fluorescein solution, while the two 10 cc beakers would each contain 10 cc of water. When the students entered class, they would see the apparatus set up as in [the figure below].

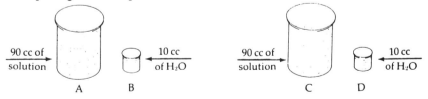

Beakers set up for solutions demo.

The students in the front rows would be able to read aloud the total number of cc of liquid in each of the four beakers, but it would be up to the instructor to reveal the nature of the ingredients and the concentration of the solution in beakers A and C.

Mrs. Armstrong intended to pour the water in beaker B into the solution in beaker A but leave the other two beakers as is, so that students might have a ready reference as to the "before" situation. She knew she must obtain answers to two major questions: (1) What changed? and (2) What remained the same? She would have to use contingency questions to elicit the hidden constant—that is, the constant volume of the fluorescein dye in beaker A both before and after the demonstration. Then she would help the students compose a list of varied practical situations where mixing of this kind might occur but where the desired final concentration would be known (in order to clarify where the class was headed).

If Mrs. Armstrong implements the above idea successfully, she should next provide students with a model problem. The model problem should be matched to the action illustrated by the demo (that is, one ingredient should be added to a mixture) although the unknown might differ. Then she will need to move gradually from the demo to the mathematical model in at least three steps (pictorial, English sentence, mathematical sentence). Mrs. Armstrong would probably cue students to help complete:

1. A labeled sketch of the situation in the case of the model problem with perhaps a "before" and "after" version.

2. An English sentence that describes what stays the same (for example, the cc of fluorescein dye "before" is the same as the cc of fluorescein dye "after").

3. The corresponding mathematical sentence with the appropriate designation of the unknown.

Notice the levels of translation that are provided for in this lesson outline. The situation and its dynamics are pictured and labeled. The importance of this kind of iconic representation of the problem is unparalleled in helping students identify the changing and unchanging quantities in the problem. Now the students must create an English sentence based on their analysis of the iconic representation. This step transforms the problem into a statement that can be translated from left to right with the first level of translation skills.

Although the new rule has been obtained at the equation-stating step, the students should still be asked to complete the problem and check the result in the statement of the problem as well as in the original equation. Moreover, as in all problem-solving instruction, it is vital that the teacher and the students review the process which led to this rule. Next comes student practice of the new rule, right? Yes, but it takes a relatively long time to completely work through even a small number of these problems. And leaving the bulk of them for homework is never a sound substitute for supervised practice.

Implementing the NCTM Standards
© 1994, Janson Publications, Inc.

Selection 2

Parabola—Wax Paper Folding

A. Although this activity can be completed by one person, it will be more effective if it is completed by several students who can then discuss the analysis questions. It is appropriate for either novice or more experienced teachers of mathematics.

B. Each person needs a sheet of wax paper approximately 30 cm on each edge. Draw a straight line, *l*, about 5 cm from one edge. Then place a point, *P*, about 8 cm from the same edge and halfway from the sides of the sheet (Figure 1). Label the point and the drawn line. Map *P* to any point on *l*, fold and then open the paper. (Be sure the fold, or crease, is evident.) Next map *P* to a different point on *l* and again fold and open the paper. Continue the mapping, folding and opening steps and look for a pattern (Figure 2).

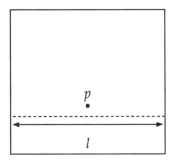

Figure 1: First step in parabola construction.

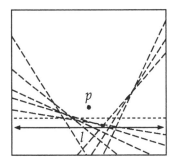

Figure 2: Wax paper parabola.

C. Respond to the questions in this section individually and then talk about your answers with others doing the activity.

  1. What kind of curve seems to have been constructed by this process of mapping a single fixed point to a set of other points along a fixed line? Explain why the construction of lines of reflection in this manner should yield this curve.

2. Reflect on your thinking as you completed three folds, then four.....When did you decide that the pattern being constructed was a representation of a parabola? Did you have any early ideas that were shown to be incorrect after further investigation? In what ways does this construction with all its folds help to make visual various properties of the parabola? You've studied the parabola before. If this was a novel experience for you, in what way(s) did it clarify or strengthen your concept of parabola?

D. You were literally constructing a parabolic representation in this activity. If you were following the directions in B mechanically, you were only interacting with materials. If you were thinking about the emerging pattern, making tentative conjectures and testing them, you were mentally active and there was a dynamic interplay between your mind and the activity. Then, when you were asked to reflect on your earlier thoughts, you were mentally active at another level, one that researchers call the "metacognitive" level—thinking about your own thinking.

Selection 3

Mr. Potter's Lesson on Solving Novel Problems

[A] teacher who wants to teach future problem-solving lessons should incorporate into rule lessons such behaviors as descriptions of the way the teacher has fumbled toward a solution, questions as to the advantages and disadvantages of various modes of attack, and hints as to ways to get feedback without recourse to the teacher.

Mr. Potter had been gradually moving his geometry class toward problem solving. He included all the above behaviors and, in addition, encouraged students to work in groups and to defend their work to one another. On the day of the first problem-solving lesson, here's what he told the class:

> We're going to try a little different approach today. You're going to behave like working geometers—conjecturing, deducing, and assessing different approaches. I've prepared a handout containing five geometry problems, all of which can be proved in more than one way. I've divided you into groups and assigned two of the five exercises to each group. When I give the signal, the groups will move to their assigned locations and the group leader will take over. The leader must see that all directions are followed and should try to give all students in the group an opportunity to participate. The group should find as many different proofs for each of the two exercises as possible, should work each out and then discuss the advantages and disadvantages of each. Is one proof more rigorous, easier to follow…? Keep a record of your conclusions so that you can report on them when I ask you to stop the group work. Questions? [Two or three questions.] When I turn on the overhead projector, you will see the location where you are to sit and the exercises assigned to your group. The leader's name is starred [overhead projector turned on]. You may move into groups now.

Mr. Potter's success today will depend partly on his prior developmental work and partly on the nature of those geometry exercises. They must be above level III in the cognitive taxonomy with enough of the known so that the students can begin hypothesizing, and with the right mix of novelty to insure the blocking of "pat" solutions. Mr. Potter's quoted preamble would constitute his introduction to the activity. You should be able to identify the next instructional strategy he would like in his plan and the associated feedback strategies. That's right. Next would come small-group discussions with Mr. Potter touring, listening for conjectures, trying to get students to help one another, praising an insightful idea, and so on.

Selection 4

Primes Lesson Plan

Topic: *Concepts of Prime and Composite*   Date: *October 4, (Wednesday)*
Class/Period: *Math 7, Period 3*

Instructional Objectives:
1. Operationally define prime and composite in terms of checker piles.
2. Define prime and composite in terms of factors.
3. Characterize natural numbers, not used in today's class, either prime or composite, and justify the characterization.
4. Give examples of prime or composite numbers different from those used in lab.
5. Use the words "prime" and "composite" in written and oral work.

Routines:
1. Take attendance, via seating chart.
2. Collect late work from Mary and Alf.

CONTENT ITEMS	SPECIAL MATERIAL	INSTRUCTIONAL STRATEGIES	FEEDBACK STRATEGIES		TIME EST.
			GIVE	GET	
Recall of concepts: even, odd, factor and principle of divisibility		T has each student count off and tells them to remember their number names. Tells Ss to stand and remain standing as "name" is called. T calls: "2, 8, 12, 6, 4,…" Asks Ss who think they know the rule to raise their hands. T elicits operational def. of even and why. T then asks for description of names of the seated Ss to get def. of odd. T uses Q/A to elicit recall of concept of factor and principle of divisibility.	T waits. T asks for more hands T takes ans.; asks other Ss if they agree	T praises Ss with hands up T confirms	5 min.
Number patterns history and informing Ss of outcomes		T reminds Ss of past lectures on interest of Greeks in number patterns and how they used pebbles to make shapes. So 3 was a triangular number; 4, a square number. T tells Ss that today they will do a lab to discover some other number characteristics first noticed by the Greeks			1 min.

CONTENT ITEMS	SPECIAL MATERIAL	INSTRUCTIONAL STRATEGIES	FEEDBACK STRATEGIES		TIME EST.
			GIVE	GET	
	lab sheets, demo set of checkers, OH (overhead) lab sheet, Boxes of checkers	T has Ss pass out lab sheets and tells all to fill in heading. T gives lab instructions: a. Ss will work in pairs as yesterday b. Each pair to get box of checkers c. Lab—to model task T will demo, to talk about patterns they find, and to record results and patterns on sheet. d. Collection, use and return of materials. T asks S to read headings on lab sheet. "We'll work together on 6. We'll need to do this systematically. We can make one pile of checkers using all the checkers." T demos and records 1 under "pile" heading and 6 under "number of checkers" heading on OH lab sheet. T tells Ss to copy these data on the same place on their sheets. T asks if all 6 checkers can be used to make two equally tall piles. T demos and records data on OH lab sheet and tells Ss to copy on their sheets. Then continues Q for 3 equally tall piles, 4, 5, and 6. T gives signal for preassigned pairs to begin lab.	T scans papers of a few. T takes ans. from one S; takes straw poll. Spreads Q	T corrects as needed T or S confirms by demo	5-7 min.
Number patterns		Pairs of Ss work on lab. T gives time signal, reminds pairs who are finished to look over data for patterns and to discuss their ideas with each other.	T tours & asks Qs on procedures and results T cks. data on sheets	S to check ea. other's strategies & results T praises progress	10-15 min.
Primes Composites checker piles Primes Composites Factor Exclusion of 1		T—Q different pairs for data to fill in on OH lab sheet. T elicits patterns from as many pairs as possible; notes common patterns and, if not elicited, directs Ss' attention to differences in "amount" and kind of data for numbers such as 13, 7, 2 vs. 15, 8, 6. If needed, uses contingency Q and repeats checker demo to highlight differences. T elicits "checker" def. of these number groups; asks for other numbers that fit one or the other group. If no S has yet described these in terms of factors, T elicits definition of these in terms of factors. T emphasizes the unique role of the number 1.	T sees if Ss agree Some straw polls. T calls on diverse Ss	T confirms T writes patterns on board Has S confirm	10 min.

Implementing the NCTM Standards

| CONTENT ITEMS | SPECIAL MATERIAL | INSTRUCTIONAL STRATEGIES | FEEDBACK STRATEGIES | | TIME EST. |
			GIVE	GET	
Vocabulary prime composite		T says and writes: *prime number* over set of examples and *composite number* over another set of examples as T asks: "What kind of a number is 3? 7? 9? 2? 14? ...? T directs Ss to copy labels and examples on their lab sheets.	T checks sample papers. T spreads Q.	T gives non-verbal + signal. T corrects as needed	3 min.
generalizing to other cases		If no S has named a number greater than those on the lab sheet, T asks whether 27 (32, 41,...) are prime or composite. Tells Ss to write ans. and reasons on back of lab sheet.	T tours and Q Ss.	Has Ss demo with checkers	3 min
Look ahead	Puzzle sheets	T hands out primes puzzle sheet to begin now and complete by tomorrow.	T tours		0–3 min.

Figure 1: Sample concept and vocabulary plan

NAME: _____ DATE:_____

LOOK FOR PATTERNS WITH CHECKER PILES

Number	How many piles equally tall?	How many checkers in each pile?
2		
3		
4		
5		
6	1 2 3 6	6 3 2 1
7		
⋮	⋮	⋮
25		

Figure 2: Part of lab sheet on primes and composites.

Teachers who have used these three dimensional models in this lab with systematic recording of data on a lab sheet have found that most students isolate the examples of the primes from the examples of the composites. If that doesn't happen in this class, the teacher has a backup strategy. Did you notice that the teacher doesn't expect to get the typical definition of primes and composites immediately? A checker definition is quite acceptable! In fact, the cooperating teacher, Mrs. Lopez, advised this student teacher not to ignore the checker definition if some student first described the two groups of numbers in terms of their factors. In her experience, some students needed to return mentally to the checker piles and she had found it was helpful to maintain the connection between that physical model and the mathematical concept. Moreover, Mrs. Lopez had another good reason; a checker lab would be used later in the week to develop the concepts of common factor and greatest common factor.

If you followed our instructions and didn't read the feedback strategies columns, now is the time to return to Figure 1 and analyze those strategies in terms of the instructional strategies opposite them. When and how does the teacher plan to get feedback on the extent to which students' thinking is starting to focus on the essential attributes of the concepts? Developmental Q/A, used in conjunction with demo and lab modes, certainly is the key to this crucial aspect. Will it work? Yes, *if* the teacher implements it well.

Selection 5

Kite Lesson

Generalizing is a valuable tool of the scientist, as is idealizing, but the processes most characteristic of the mathematician are those which result in a mathematical system. Somehow a study of Euclid's geometry only reveals part of the picture. Until students get their own hands dirty in the task of constructing an axiomatic chain, they have only partially learned these basic deductive processes. An exercise within the grasp of geometry students who have studied the family of parallelograms and learned how to deduce the properties of each member of the family centers around a quadrilateral called a "kite." A kite is defined as a quadrilateral with two pairs of congruent adjacent sides. *ABCD* is one example. The students are asked to deduce the properties of a kite and state these as theorems and corollaries. They may define special kites. For example, a "right" kite could be defined as [a] kite with a right angle included by a pair of congruent sides, as in *MNOP*. Of course, it could also be defined in other ways. Here is another opportunity for student discussion of alternative choices for definitions and the usefulness of each. Most important, the students have to work at the deduction process. Unlike the typical sequence of theorems involving the parallelogram family, this sequence is not familiar to students. They have no preconceived notions of the form of theorems further down in the chain. They actually have to deduce these from the definitions on which they've agreed. This kind of thinking actually takes more than one 40 or 45 minute lesson. The teacher can start the class on the work, but would be well advised to design a strategy in which small groups begin to work on the chain of theorems in class. Reports could be given after two or three days of work on the project outside of class. The kite exercise can do much to dispel the mystique of axiomatic thinking (Farrell, 1970).

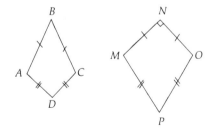

Selection 6

Sample Projects

Here are two examples of project directions used in senior high school mathematics classes.

1. Math 10 Project   Date: February 8

    A. Choose one of the following topics:
    Linkages; Space-filling; Curve stitching; Polyhedral decorations

    B. Design an original model. check out the mathematics of your design with the teacher. Then build a durable, attractive model and be prepared to explain its properties to the class.

    C. Submit your plan on a five-by-eight-inch index card by February 15. Completed projects are due February 22, and class time will be reserved for reports the week of February 25.

2. Math 12 Project   Date: October 1

    A. Choose *one* of the following topics:
    Algebraic Balancing of Chemical Equations
    Algebraic Solution to Electrical Circuits
    Minimal Surface Area Experiments with Soap Films
    Numerology
    Ciphers, Codes, and the Way They Are Broken
    Mathematical Aspects of Population Growth
    Statistical Study of Finger Length Variation in Adolescent Hands
    Mathematics in Sound and Music
    Function Before Fashion—Mathematical Designs
    Timetables, Calendars, and Clocks
    Analysis of Hurricane Paths

    B. Research the topic and prepare an 8- to 10-page typed paper on the results of your findings. Include a bibliography of at least four references. You may *not* include an encyclopedia as one of the four sources! Be sure to cite these references in the format given in class.

    C. Submit a one-page outline by October 15.

    D. Completed papers are due November 1.

Implementing the NCTM Standards
© 1994, Janson Publications, Inc.

If students complain of the extra work involved in a project, then the teacher has goofed somewhere along the line. The project should be threaded into the coursework. It is up to the teacher to design assignments and daily work so that students see this mode as an alternative, but different, way of learning relevant aspects of mathematics. The students soon learn that the teacher doesn't mean what is said if no class time is reserved for reports, or if the end of the project means little more than a grade in the teacher's record book.

Selection 7

A Lesson in Networks

Just as the search for generalization is an aid to problem solving, so the process labeled idealizing in the model of mathematics represents ways to simplify a sometimes unwieldy problem. The solution of a well-known problem, called the Königsberg bridge problem, is a good example of idealizing in action.

In the eighteenth century strollers in the German university town of Königsberg walked along the shores of the Preger River and over the seven bridges which connected two islands to each other and to the mainland (see Figure 1).

The problem which one of these strollers is said to have posed was in the form of a question: "How can you take a walk so that you cross each of our seven bridges exactly once?" The problem was eventually solved by Leonhard Euler, who idealized the situation in the following manner. The bridges will be drawn as line segments; the islands and shore, as points of intersection (see Figure 2). Try to solve the problem yourself before reading further.

In the process, Euler invented networks, an aspect of that branch of mathematics called topology. Euler's analysis of networks led him to characterize points of intersection as either "odd" or "even vertices." *A* is an odd vertex since 3 arcs intersect at *A*. There are no even vertices in the Königsberg bridge network. He found that the number of odd vertices is limited to two or less if the network is to be traveled without retracing any arcs. So Euler's answer was, "It can't be done!" Euler's use of idealizing processes eliminated extraneous data which had only confused others who attempted to solve the problem.

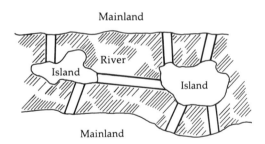

Figure 1: Königsberg bridge diagram.

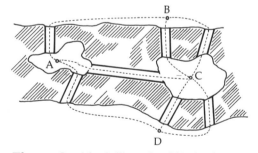

Figure 2: Modelling the Königsberg problem.

Selection 8

Table on Everyday Ideas

Everyday ideas	Mathematics questions
1. Statistics on deaths by car accident and/or motorcycle accident involving teenage drivers over a particular holiday weekend (see *Time* or *Newsweek* or check with National Safety Council).	1. Over the past three years, what are the trends in type of accident (graphs, ratio)? Predict the number of deaths per type of accident for this year, for each holiday. What factors might alter the chances of your predictions occurring?
2. Obtain a completed scoring sheet from a bowling game.	2. Write a flow chart that would tell a non-bowler exactly how to compute a friend's score.
3. Obtain a city map that can be clearly projected. Mark the well-known locations (school, churches, city hall, stadium, city park, etc.)	3. Have students give the coordinates of specified locations, using map indices. Have them locate their homes and name the coordinates. Why are such maps usually indexed by letter and number, rather than by two numbers?
4. My size, my height, my weight—how do I compare with my classmates? A young child?	4. Estimation uses are unlimited. Make out a personal chart with your height in meters, the length of your fingernail in cm, the length of your shoe in cm, the distance from your elbow to the tips of your fingers in cm, and your weight in kg. Compare the weight of a paper clip and a penny in g. Use your table to estimate the dimensions of your desk, the height of the door, the dimensions of the hockey rink, and the weight of a softball, football, or basketball.

Ms. Blumenstalk's Area Proofs

Ms. Blumenstalk approached the sequence of area proofs in geometry by having her tenth grade students estimate the area of irregular shapes in a lab activity. They covered the shapes with different congruent real world objects from pennies to bananas. Then the students used some of the more interesting Escher patterns, perhaps a tile like the ones in [the figure below]. They worked in groups and discussed the advantages and disadvantages of the different shapes as area unit figures. Eventually, they worked on grids, triangular and parallelogram, as well as square grids and came to an understanding of the usefulness of square grids. When Ms. Blumenstalk turned to a consideration of the proofs for the well-known area formulas, she found that these tenth graders had a new appreciation of the relationship of those formulas to the area concept and possessed the concrete prerequisites to suggest transformations of the figures to get equivalent areas.

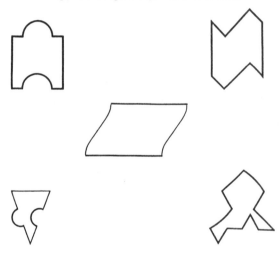

Modeling Common Algebraic Identities

The Greek use of regions is not just a historical curiosity, but also an excellent way to introduce some of the algebraic formulas we teach. Have students use cutout square and rectangular shapes and attempt to generate different ways of picturing the same algebraic phrase. You will have to define ($a+b$) as the placing of two strips of differing lengths next to each other, while (a-b), where a is longer than b, is shown by overlapping the strips (see Figures 1a and 1b).

Next, demonstrate to them the product of two quantities as a rectangular region whose dimensions are the two factors, as in Figure 2. Students will easily obtain the usual equivalent for $a(b+c)$, but will find that transforming a^2-b^2 requires more ingenuity (see Figure 3). Remind them that while you (and perhaps some of them) used algebraic tools to obtain the equivalent results, the Greeks obtained these by testing and merely using their agreed-upon assumptions. They used a result if it appeared to be helpful or simpler. Your students may obtain results that the Greeks chose to ignore. For example, $a^2-b^2=(a-b)^2+2\cdot b(a-b)$. Of course, this expression can be transformed into the typical $(a+b)(ab)$ by dividing by the common factor (ab). Would you expect the average ninth grader to understand that transformation? Reconsider the research results on reasoning, especially reasoning in algebra from Chapter 3. So what can the teacher do? Here is an instance where the physical transformation of cut-out shapes helps students get a useful result in a way they can understand. They can be instructed to try to form single rectangles or squares, rather than L-shaped figures, whenever possible, since then only one area needs to be represented.

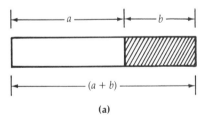

(a) (b)

Figure 1

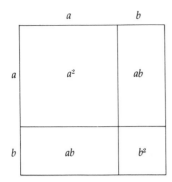

Figure 2

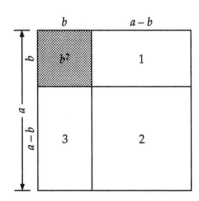
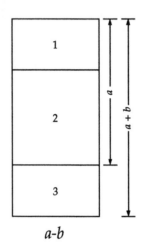

Figure 3

Implementing the NCTM Standards
© 1994, Janson Publications, Inc.

Selection 11

Probability Lab

T begins with a ten-minute lecture with demo related to finding the empirical probability of an event. A three minute Q/A session, in which T sampled eight students, provided positive feedback on Ss understanding of (1) the empirical techniques used to obtain number of favorable ways the event could happen, (2) the "sample space" method used to record data, (3) the location of the various games of chance and problem cards, and (4) the roles of each lab partner.

T then gives the signal to begin work and takes a position in the room from which the Ss collection of materials and equipment can be observed. All Ss gather materials quickly and quietly, then return to their assigned laboratory places. At this point, T begins to move quickly and systematically from group to group. T notes that the first four groups visited are proceeding exactly as directed but just as T [reaches] the fifth group, loud voices [are] heard coming from the opposite corner of the room.

T moves quickly to the loud group, touches one S on the shoulder, and uses the finger-across-the-lips signal when the S looks up. Both partners quiet immediately and T asks if they are having a problem with the exercise. Both partners start to talk at once, each complaining that the other is misinterpreting the card problem. (The problem is: What is the probability that two jacks will be drawn from a standard deck of 52 cards, when two cards are drawn without replacement?) "Herman is drawing the cards one at a time," complains Yvonne. T holds up one hand in a "stop" gesture; the Ss quiet down; and T uses Q/A with each in turn to clarify the procedure. The partners return to work; T observes they are now working amiably, and T moves on to other groups that are all proceeding in an orderly fashion.

As T continues to move from group to group, observations indicate that all Ss are now interpreting the directions correctly in the case of the coin, dice, card, marbles, and the roulette wheel events. However, the first three groups to progress as far as the event using the roulette wheel and the spinner have difficulty agreeing on the nature of the events that are to occur. T turns off the lights for three seconds. All the Ss quickly stop whatever they're doing and look at T, who then turns the lights back on and gives a lecture-demo on these particular events. All groups then return to work at T's signal, and T's subsequent observations while touring groups indicate all are now obtaining appropriate data.

With about five minutes of work time left in the class period, T [overhears] the following conversation [between] one pair of partners:

Saul: "We calculated the probability of a black number showing on the roulette wheel as 42/100 and of a 2 occurring on a spin as 3/100.

If we want the probability of both happening at the same time, we would have a combined probability equal to the product of 42/100 and 3/100."

Alice: *"No, we wouldn't! We would just be adding one event to another, so the total probability would be the sum of 42/100 and 3/100."*

Saul: *"Look, think of it this way. The chance that both events will occur at the same time is less likely than for just one to occur. A sum would signify just the opposite."*

T praises Saul and Alice privately for thinking ahead and gives each credit for having a certain amount of logic. A short Q/A sequence by T gets Saul and Alice to compare this problem with the card problem where two jacks had to be drawn. T asks both to rethink the definition of probability based on sample space usage. Then T goes to the chalkboard, writes a brief statement of the problem posed by Saul and Alice, and for the evening's assignment asks each student to write out a sample space using ordered pairs to solve this problem. Then each student is to arrive at a generalization for the $P(A \cap B)$ and to obtain the probability of the first event or the second happening, given the same data. A brief flickering of the lights signals "clean-up," which progresses smoothly and quickly. T calls attention to the assignment, publicly gives credit to Saul and Alice for thinking ahead, and praises all for their very good work during the period. The bell rings, the Ss leave, and T checks to see that all is in readiness for the next class, which is to begin in four minutes.

Note the highly positive feedback the teacher received on the effectiveness of the introduction to this laboratory exercise. Identify all the things the teacher did to ensure that all students wold begin the laboratory work effectively and efficiently. Also note how the teacher found out that only one group was encountering difficulty with the card problem and how this problem was handled without interrupting the other students who were proceeding without problems. Nonverbal feedback giving, as well as feedback getting, is highly effective for all who learn how to use it.

Apparently the teacher was quick to read the negative feedback collected on the roulette wheel/spinner problem and made use of it to reorient the entire class. If the teacher had failed to interpret this as a widespread problem, no doubt it would have been necessary to treat the same problem over and over in successive individual visits to nearly all the groups.

The student-to-student discussion between Alice and Saul not only provided the teacher with valuable feedback, but also presented a golden opportunity to reinforce prior learning on the nature of the mathematical enterprise and to provide for an in-context extension of that learning by means of a relevant homework assignment.

Selection 12

Bicycle Gears Lesson

Ratio and proportion could also be introduced by studying the relationship between the number of teeth on gear arrangements and the revolutions of the gears. Why not start with an example from the world of many teenagers, a bike with gears? We recommend that you locate a three-speed bike, or use the first, intermediate and highest gears on a bike with ten or more gears. Why does first gear allow you to pedal with relative ease when biking up a steep hill? Turn the bike upside down and have one student turn the pedals. The number of revolutions of the pedals and the corresponding number of revolutions of the rear wheel need to be recorded for first, second, and third gear (lowest, intermediate, highest) respectively. Does the ratio change if the rider goes at a faster pace? Why or why not? If you're working with a junior high class, a simple laboratory with cardboard gears would be a good way to help them understand this application of proportion. [The following figure] depicts models of two different gears (30 teeth and 15 teeth), which may be copied and traced on sturdy cardboard, such as the backs of tablets. After the models have been cut out, they should be fastened to a cardboard backing by means of paper fasteners through the center of each gear. The students move the gears so that the starting points touch and then [the students] move the larger gear through a complete revolution while counting the number of revolutions of the smaller gear. A data table is completed (an example follows) and the students are asked to predict the number of revolutions of the smaller gear for 5, 6, and 12 revolutions of the larger gear.

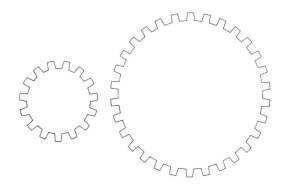

Gear models.

Data Table for Gear Laboratory

Gear A No. of turns	Gear B No. of turns	No. turns A / No. turns B
1	2	1/2
2	4	2/4=1/2
3	—	—
4	—	—

The students are thus guided to compute the ratio of the number of teeth of the two gears, perhaps *before* the term *ratio* is known. There are one-half as many teeth in gear B as in gear A. What would happen if a still smaller gear with 10 teeth were meshed with the larger gear? An extension of the laboratory would be a fine way to explore the students' conjectures; or, if time became a problem, a demonstration on the overhead projector could be used to answer the question. Spirograph gears make excellent overhead projector models for this purpose as long as the teacher first selects gears with numbers (of teeth) that are factors of one another.

Implementing the NCTM Standards
© 1994, Janson Publications, Inc.

Selection 13

Mr. Greenberg's Exponent Lesson

Unlike inductive reasoning, deductive reasoning *is* the method of proof, but it has its own constraints, which are important facets of this mathematical process. Mr. Greenberg's explanation of a^0 to his senior high class contains a subtle error which clouds a major process used by mathematicians. See if you can find it.

> *(The class had used the rules for $a^m \times a^n = a^{m+n}$ and $a^m/a^n = a^{m-n}$ where $m > n$ and where m and n are natural numbers.)*

> *Mr. Greenberg: "If $a \neq 0$, then $a^6/a^6 = 1$ since any nonzero quantity divided by itself is 1. But using our rule for division of like bases, $a^6/a^6 = a^{6-6} = a^0$. So, since quantities equal to the same quantity are equal to each other, $a^0 = 1$."*

Have you found Mr. Greenberg's error? That's right—he ignored one part of the assumption for using the division rule. He missed a great opportunity to illustrate the way mathematicians alter assumptions and construct definitions so that rules may be extended. Here *m* is equal to *n* and a zero exponent will result from application of the division rule. Hence the mathematician makes a *tentative* assumption that *m* and *n* may be whole numbers with $m \geq n$ in the case of the division law. Then, the mathematician checks to see if that assumption will lead to contradictions. In this case, no contradiction was found so the next step, formally defining $a^0 = 1$ if $a \neq 0$, was taken. Finally, it is proved that the extended laws hold under the new definition.

Selection 14

Mr. Ramirez' Lesson in Introductory Algebra

Why is the product of two negative numbers positive? If you were to survey algebra teachers on the number of times this question is asked, you would hear a resounding, "Too often!" One of the most straightforward ways to approach the topic of multiplication of directed numbers, and thus emphasize the way definitions are constructed, is through a careful developmental lesson in which pattern-analysis is stressed.

First, the teacher must establish that it is reasonable to expect the multiplication of two positives to be positive. Students typically readily agree to this. Then the teacher needs to emphasize that since these directed numbers represent both quantity and direction, that mathematicians had to define multiplication, but wanted to construct a definition that wouldn't conflict with other mathematical patterns. Then, let students look for the patterns as they answer the following questions written, one-by-one in a carefully sequenced way on the board.

Mr. Ramirez writes

$$\begin{array}{r} {}^+10 \\ \times \, {}^+5 \\ \hline \end{array}$$

"By our agreed-on definition, the product must be what?"

Then, next to that:

$$\begin{array}{rr} {}^+10 & {}^+10 \\ \times \, {}^+5 & \times \, {}^+4 \\ \hline \end{array}$$

"Be watching for patterns. Don't answer too soon. What is the product?"

Mr. Ramirez continues in this fashion, getting products, but waiting for more and more students to indicate that they see one or more patterns. He is careful to write the positive sign in every case. After the fourth example, he asks students for all the patterns they think he is using to write the examples. They notice that the multiplicand is the constant $^+10$, that the multiplier is decreasing by one each time and that the product is decreasing by ten each time. Finally, the array looks like this:

Find the products

$$\begin{array}{cccccc} {}^+10 & {}^+10 & {}^+10 & {}^+10 & {}^+10 & {}^+10 \\ \times \, {}^+5 & \times \, {}^+4 & \times \, {}^+3 & \times \, {}^+2 & \times \, {}^+1 & \times \, \underline{} \\ \hline {}^+50 & {}^+40 & {}^+30 & {}^+20 & {}^+10 & \end{array}$$

Implementing the NCTM Standards
© 1994, Janson Publications, Inc.

"What do you think I will write next?"

Mr. Ramirez gets the response of zero for the multiplier and then reinforces the meaning of constructing definitions for this new set of numbers.

"Now we haven't defined multiplication for zero times a signed number, but what should the product be if the pattern of the products is to be maintained?"

After all agree that zero is the only number that will maintain the pattern, Mr. Ramirez emphasizes that, in general, zero times any signed number is zero, by definition—a definition that keeps intact patterns such as this one. The next step is to extend the three patterns with the help of the class. The entire sequence now looks like this

+10	+10	+10	+10	+10	+10	+10
× +5	× +4	× +3	× +2	× +1	× 0	× 1
+50	+40	+30	+20	+10	0	

"Now we haven't defined the product of a negative and a positive, but if the pattern of the products is to hold, what must the missing product be?"

Concrete operational students can reason inductively and, given enough instances and careful attention to all of the patterns, will respond with the answer ⁻10. For practice in extending the pattern, the teacher should continue the sequence of questions for three or four more examples. This helps to reinforce the new definition. Next, Mr. Ramirez needs to introduce the need for the commutative property. Students will generally agree that it would be convenient if this property held for multiplication of these signed numbers. It's important for them to realize that mathematicians think about convenience and ways to construct simpler rules or ways to adapt an existing rule so it holds for more situations.

After some oral practice on multiplication restricted to two positives or to a positive and a negative, Mr. Ramirez is ready to use a similar technique to lead the students to the construction of the definition of the product of two negative numbers. The sequence might be built up, with questions very similar to those used earlier and the instruction to again look for all patterns.

Find the products

10	10	10	10	10	10	10	10
× +5	× +4	× +3	× +2	× +1	× 0	× ⁻1	×
⁻50	⁻40	⁻30	⁻20	⁻10	0	?	

Again the zero product question can be asked. Mr. Ramirez emphasizes the horizontal patterns that the students have identified and asks the question:

What number must take the place of the ? for the pattern of products to be maintained?"

Now that the students have confidence in the power of patterns, most, if not all, students agree that $^+10$ is the needed replacement. Further extension of the sequence of examples again helps to reinforce the pattern and allows the students, or the teacher, to verbalize this definition, that the product of two negative numbers is a positive number.

This particular classroom-tested presentation is effective in helping students to construct the needed definitions and is an excellent way to make real one of the products of mathematics—the definition— and the very dynamic way in which those definitions might have been constructed by mathematicians.

Selection 15

Mr. Cain's π Lesson

What can we do when the students just haven't enough mathematical background to even informally defend a generalization? Notice how Mr. Cain handles the finale in his seventh-grade laboratory lesson on π.

The students have measured the circumference of various cans with string and the diameter with metric rulers. All groups have found *C/d* to the hundredths place on their calculators.

"The group leaders should come to the board and fill in the data table."

After all have completed recording their results, Mr. Cain and the class begin to talk about the pattern they see in the resulting computation.

Mr. Cain: *"We noticed that all groups obtained a result of about 3.14. Why do you think there were some differences in the hundredths place?"*

Mark: *"It was hard to use the string and sometimes we were careless."*

Rosie: *"Some of the cans had rough spots on the top. Maybe those bumps caused trouble?"*

Mr. Cain: *"That's very good! Measurement inaccuracies are always with us. In fact, even high-powered scientific tools have measurement inaccuracies. Measurement is never exact. So we don't prove anything by using measurement, but we do use measurement to test ideas. As a matter of fact, your data seem to show that C/d is always a little more than 3. Would measuring more cans, bigger and smaller ones, prove that?"*

Johnny: *"No, but I bet we'd get the same thing."*

Mr. Cain: *"Aah, you're saying it seems very convincing! You're right. More data are what the scientist would need to test that pattern. But the mathematician always wants to prove rules. Do you know that centuries ago mathematicians did prove that, for every circle, C/d is always exactly the same number? They used an argument you'll study in high school. They also found out that the number they obtained was different from the ones we get when we measure objects. So later mathematicians gave that number a new name. They called it π (pronounced "pi"), a letter of the Greek alphabet. But they did want to measure circular objects, just as we do. Think of the circular shapes found in sports alone—the markings on the basketball court, the cross-sections of baseballs and golf balls, and the outline of an archery target. So we use an* *approximation of a number we can measure with, a number you obtained in your lab, 3.14. Sometimes we'll use 22/7 as an approximation of π. That's sometimes easier to work with, but it's still only an estimate. Now let's see how to use this famous number."*

Mr. Cain will have to return to these ideas often, because the concept of π is difficult to grasp. Why didn't he introduce the label *irrational*? Did you notice that he never stressed the difference between the physical models of the circle in the can shapes and the mathematical model that they represent? He'll have to be alert in follow-up lessons for the appearance of a common misconception, that 22/7 actually equals π. Since 22/7 is approximately equal to 3.142, some mathematics students believe that a computer program in which 22 is divided by 7 will result in the decimal expansion of π. "But π is irrational!" we hear you cry. Mr. Cain and his senior high colleagues have a lot of spiraling to do in helping students understand this concept.

Selection 16

An Algebra Lesson

Consider an exercise based on a generalization obtained from a powerful mathematical product, Pascal's triangle.

$$1$$
$$1\ \ 1$$
$$1\ \ 2\ \ 1$$
$$1\ \ 3\ \ 3\ \ 1$$
$$1\ \ 4\ \ 6\ \ 4\ \ 1$$
$$1\ \ 5\ \ 10\ \ 10\ \ 5\ \ 1$$

Ask the student to add the numbers in each row of Pascal's triangle.

$$1 \qquad\qquad = 1$$
$$1+1 \qquad\quad = 2$$
$$1+2+1 \qquad = 4$$
$$1+3+3+1 \quad\ = 8$$
$$1+4+6+4+1 \ = 16$$

Then ask them to find, without adding, the sum of the numbers in the 15th row, the 24th row, the nth row. Next challenge them to continue the following number pattern and compare results with that obtained in summing the numbers in the rows of Pascal's triangle.

$$11^1 = 11 \qquad \text{and} \qquad 1+1 \qquad\ \ = 2$$
$$11^2 = 121 \qquad \text{and} \qquad 1+2+1 \qquad = 4$$
$$11^3 = 1331 \qquad \text{and} \qquad 1+3+3+1 \quad\ = 8$$
$$11^4 =$$

This apparent generalization eventually fails. It is important that the students are not told ahead of time that, in this case, the pattern will break down. A few examples like this mixed in with others where generalizing does yield a valid result will suffice to warn the students that care must be taken in the use of data.

The same illustration pinpoints the second danger in the use of generalization. When students fail to verify a presumed generalization, they fall into the trap of abusing this inductive process. Results must be checked against more data and, in mathematics, proved by deductive processes. Stop right here and test yourself. In the first example, what generalization did you obtain for the nth row?...Then return to the second example and compare the data from this exercise with that from the first. Try to pinpoint the arithmetic behind the lack of a continuous, common pattern.

Selection 17

Ms. Pauli's Lesson on Decimals

Inductive reasoning is an excellent way to help students see for themselves some of the rules and concepts of mathematics. Sometimes we call it *looking for patterns.* However, inductive reasoning must be treated with care. In the transcript of Ms. Pauli's lesson, several excellent tactics are employed, but she makes at least one serious error. Identify both the positive approaches she used and the error. Ms. Pauli is trying to have her junior high students "see" a pattern—in this case, a rule for multiplying two decimals.

"Class, as I write each example on the board, raise your hand if you know the answer."

1. $2/10 \times 3/10 = ?$
2. $5/10 \times 4/10 = ?$
3. $7/10 \times 6/10 = ?$

Ms. Pauli quickly got answers of 6/100, 20/100, and 42/100. Next she wrote the following examples on the board and asked the class to compare these with the first three examples.

4. $.2 \times .3 = ?$
5. $.5 \times .4 = ?$
6. $.7 \times .6 = ?$

"Cindy, do you have an idea?"

"I know the answers. They're the same as before."

(Many hands wave agreement. Joe says that all Ms. Pauli did was change the fractions in the first three examples into a decimal form.)

"That's very good. Now try the next three examples."

7. $.9 \times .2 = ?$
8. $.8 \times .8 = ?$
9. $.3 \times .5 = ?$

After checking the answers, Bob says that he has found a shortcut.

"I just multiplied the whole numbers, 9×2 in example 7, to get 18. Then I counted to the left for two decimal places."

"Class, let's test Bob's rule. Does it work for example 8? 9?"

(All agree Bob's rule works for these examples.)

"Good. Let's test it on examples 4, 5, and 6."

(The class takes more time over example 4, but finally all agree that Bob's rule works for these three examples, too.)

"Now, class, all of you use Bob's rule on examples 1 and 4 from page 17 in your text. If it works there, we've proved that Bob's rule is correct."

Did you notice the sequence in which Ms. Pauli presented the examples? She wanted her students to observe data, classify that into a pattern based on previous knowledge, and be able to apply that pattern. Look at the examples again and try to identify the characteristics which seem to be positive aspects of pattern constructing.

Where is the error? Ms. Pauli's class did obtain a generalization from these instances but they didn't *prove* it. Nor will they prove it by further tests on other instances. Further positive tests will simply improve the *probability* that Bob's rule is the desired one. *Inductive reasoning never results in proof.* Could Ms. Pauli have *proved* Bob's rule? Yes, quite easily, if she had merely let Bob defend his rule. He might have argued that it followed from the earlier multiplication examples and the rewriting of a fraction with a denominator of 100 as a decimal. Bob may have used different words, but if his message was a paraphrase of that explanation then he would be using deductive methods of proof. Should eighth-graders bother with proof at all? Yes, whenever possible, for the essence of mathematics is lost without it, but teachers should encourage informal arguments and avoid formalistic rigor.

In this case, there is another flaw in Ms. Pauli's lesson. The conceptual pattern that is at the basis of the multiplication of decimals was captured more clearly by Cindy and Joe who saw that fractions with denominators of 10 or 100 were simply other forms of decimals. By seeming to emphasize Bob's shortcut, instead of Cindy and Joe's pattern, Ms. Pauli may be asking for trouble. Bob's shortcut may represent a misconception—a superficial pattern based on these limited examples. If he were asked to give the answer to $.3 \times .06$, perhaps he would apply his rule blindly and get a result of .18. Ms. Pauli must be sure that neither Bob, nor any of the other students, has invented a simple rule that is not based on meaningful mathematics. Careful questioning and gradually differentiated examples are always part of a lesson that includes finding patterns. The students will invent their own strategies. The teacher has to help them check on these strategies and connect them to earlier concepts and rules.

Selection 18

Honeycomb Lesson

Let's start with an activity based on [the following questions: What shape are the compartments in a honeycomb? Would some other shape be more efficient (in terms of the survival of the bee)?]. In a junior high class, the teacher can show the students a picture of a cross-section of a honeycomb or, better yet, pass around (and project on the overhead) a portion of a real honeycomb. Don't assume that all the students have seen a honeycomb. Many supermarkets sell only jars of honey, minus honeycomb. Next the problem posed in the question has to be spelled out. What is meant by "efficient" in this case? (Here's a good place to involve a science colleague directly in the class, or indirectly as a resource person. Be sure to ask that science teacher about the two aspects of the bee's behavior: building the comb and storing the honey. Also seek your colleagues' assistance in alerting students to the dangers of thinking about the bee's behavior in terms of human attributes and reversed cause and effect relationships. Don't feel guilty about asking for this kind of assistance since your science colleagues will certainly need your help in assisting students to make proper use of mathematics in their science classes.) The "efficiency" question should eventually get restated so that quantitative techniques can be used. Students should be helped to see that there are two main issues.

1. What other shape(s) would "fill the plane"?

and

2. Given a fixed surface area to use for a cell wall, which shape provides for maximum volume of the enclosed space?

Figure 1 depicts some common shapes that could be considered as the students try to model the situation of a comb.

Figure 1: Potential "cells" for comb

Multiple congruent cutouts of each of the shapes could be used to answer question 1. The students might be asked to "tile a floor" and, as a result, to identify those shapes that make good tiles and to determine the number of such tiles needed around a point. It is a good idea to include large (such as 10 cm on an edge) and small (such as 4 cm on an edge) sets of each shape. Then, elicit from the students conclusions, such as "six equilateral triangles fit around a point whether the triangles enclose a large or a small region." A table like [the following] can be used to record data.

Shape of Tile	No. of Sides	Fills the Plane	No. of tiles around a point	Measure of one angle
Circular	NA	No	3	NA
Equilateral Triangle	3	Yes	6	60°
Square	4	Yes	4	90°
Regular pentagon	5	No	3+	108°
Regular hexagon	6			

A Partially Completed Data Table

Protractors can be made available if students have forgotten (or are just beginning to investigate) the angle measures of some of the shapes. The students can be encouraged to find patterns in the data, such as the recurring product of 360° (6 × 60°, 4 × 90°, 3 × 120°), and the increasing number of sides coupled with the decreasing number of tiles around a point. Ask the students to look at the pattern under the "No. of tiles around a point" column and explain why that pattern "tells" us to end the investigation. If the pattern of a product of 360° holds, then no wonder the regular pentagon wasn't a useful tile: 108° would have to be a factor of 360° and it isn't. Don't miss the opportunity to point out the inexactness of measurement and the differences between the cell of the comb, the cutout hexagonal shape, and the mathematical thought model, the hexagon. Finally, emphasize the processes of inductive reasoning (looking for patterns, extending patterns) and of deductive reasoning (explaining the significance of the 360° product of angles) that were being used by the students throughout this investigation.

The results of this part of the investigation make the construction of the comb even more puzzling. Surely, it would be easier to build a three- or four-walled cell rather than the six-walled cell! Perhaps the solution of the second problem we posed originally will help here. Although the students could use the same cutout shapes and simply compute areas, we've found that a laboratory activity designed to produce a three-dimensional visual solution is far more productive. You can use old manila folders (or other suitable substitutes), which you will need to cut into strips 2 cm by 12 cm. Students working in groups of two or three should be asked to construct a cross-section of a comb. Some groups should construct a comb with hexagonal cells; others, a comb with square cells; and the rest, a comb with triangular cells. In *all* cases, the surface area of the wall of any *single* cell will be the same, 2 cm × 12 cm. Each group will need a supply of strips (8 to 10), tape, a scissors, and a metric ruler. (The group constructing square cells won't need a ruler. Why not?) The Teacher needs to remind the class that combs do not contain double-walled cells. Furthermore, the teacher will need to precede this laboratory with a demonstration of the construction of a single cell and the subsequent attachment of an adjacent cell. As students work through the laboratory, and then compare results visually by placing one kind of honeycomb cell over another, the capacity-advantage of the hexagonal cell over either of the others should be unmistakable.

The "combs" in Figure 2 were produced by groups in one class. The combs were taped to the board and all the students observed the structural problem, which had been commented on earlier by several small groups. "The triangle is a rigid figure" became a meaningful statement to all the observers. (Then, why don't the hexagonal cells droop in the honeycomb? If you're not sure, check with a science colleague.)

Figure 2: Models of combs.

Selection 19

A Lesson with *The Hobbit*

Literature is another field with rich possibilities for teaching mathematics. Even before *The Hobbit* became popular, a creative mathematics student teacher (Feldt, 1977) decided to use ideas drawn from Tolkien's fantasy in order to develop applications of ratio and proportion in a ninth-grade algebra class. The students were introduced to that particular part of the adventure in which the hobbit, Bilbo Baggins, 13 dwarfs, and a wizard undertake a pilgrimage to obtain the riches belonging to the dwarf Thrain. Each student was given a copy of an adaptation of Tolkien's map in color and a set of 15 questions that developed the story line and required the application of ratio and proportion principles. An artist's version of the Feldt map (see the figure on the following page) and the first 7 questions that the students were to answer are reproduced to illustrate how the teacher wove the story line through a sequence of mathematical exercises.

1. To begin, if you are curious as to the height of Bilbo the hobbit and his dwarf companions, use the following information: The ratio of the height of Bilbo to that of the wizard Gandalf is 1:2 and the ratio of the height of Gandalf to that of the average dwarf is 4:3. How tall are Bilbo and the average dwarf if Gandalf is 6 feet tall?

2. Find the ratio of the height of Bilbo to that of the average dwarf.

3. The goal of the far-wandering group is to reach Lonely Mountain where the ancient dwarf king Thrain has left his riches. How many miles high is the highest peak of Lonely Mountain (as indicated by the flag)?

4. After a last good home-cooked dinner at Elrond's Last Homely House, the party must cross the Misty Mountains. Using the straight-line distance, how far is it from Elrond's front door through the mountains to the old ford?

5. But while crossing the tunnels of the Misty Mountains, the adventurous group is surprised by goblins—big goblins, great ugly-looking goblins! And Gandalf, the wizard, is nowhere to be found. That leaves 14 brave little souls in the party. If there are 6 goblins to every dwarf and 2 goblins for Bilbo, find the total number of goblins to be fought.

6. To find the height of an average goblin, go back to the information concerned with dwarfs in problem (1) and note that the ratio of the heights of a dwarf to that of a goblin is 6:7.

7. What is the ratio of the height of a hobbit to that of a dwarf to that of a goblin?

The journey of the hobbit.

From C. C. Feldt, "Ratios and proportions with a little help from J. R. R. Tolkien," *New York State Mathematics Teachers Journal* 27 (Winter 1977): 26-27. Reprinted by permission of the editor and author.

Selection 20

Mr. Short and Mr. Tall

Farrell and Farmer (1985) in a study involving over 900 tenth, eleventh and twelfth grade students in college-bound mathematics and/or science classes, found that approximately 47% of the students could not solve a puzzle problem involving a simple direct proportion, where the variables were related in a 2:3 ratio. The puzzle problem, which we called Mr. Tall and Mr. Short (see the figure below), was developed by Robert Karplus and used both in multiple research studies, some of which have been cited earlier, and in inservice workshop materials (Karplus, R., Lawson, A. E., Wollman, W., Appel, M., Bernoff, R., Howe, A., Rusch, J.J., & Sullivan, F., 1977, p. 1–6). The common erroneous strategy was an additive one. Such students said that because Mr. Tall was two buttons taller than Mr. Short, he should be two paper clips taller. When a random sample of the successful students were interviewed on a hands-on proportionality task involving direct as well as inverse proportions (Inhelder and Piaget's [1958] projection of shadows task), only 24% were able to demonstrate successful reasoning. They were unable to correctly apply inverse proportional reasoning. The evidence supports Piaget's contention that proportional reasoning is a late acquisition of the formal stage of reasoning. That explains the difficulty teachers report with the teaching of fractions, percents, conversion from one system of measurement to another and other diverse topics requiring the understanding of proportions.

Mr. Short

Mr. Tall and Mr. Short

The figure at the left is called Mr. Short. We used large round buttons laid side-by-side to measure Mr. Short's height, starting from the floor between his feet and going to the top of his head. His height was *four* buttons. Then we took a similar figure called Mr. Tall and measured it in the same way with the same buttons. Mr. Tall was *six* buttons high.

Now please do these things:

1. Measure the height of Mr. Short using paper clips in a chain provided. The height is __.
2. Predict the height of Mr. Tall if he were measured with the same paper clips.
3. Explain how you figured out your prediction. (You may use diagrams, words, or calculations. Please explain your steps carefully.)

Velcro Board Tangrams

If you want to manipulate shapes as a basis for developing a formula, such as the area of a parallelogram, a feltboard (or a velcro board) is useful. Cover one side of a sheet of plywood with felt or velcro. (Ask the industrial arts teacher for help.) Staple it securely to the reverse side. Use light poster paper to trace and cut out the shapes you want to move. Paste a piece of flannel or a piece of velcro on each cutout shape, depending on the type of board you've constructed. Then place the shapes on the board; move them to different positions, move them again. Some teachers use this kind of board in a classification task. First the board is separated, perhaps by two large yarn enclosures. Then the movable shapes (*e.g.* parallelograms and non-parallelograms) are moved by the students into the correct enclosure. A board of this type can by used to show the results of dissection problems, perhaps with prepared tangrams. [The figure below] shows the result of rearranging a tangram set to cover a given shape. We suggest that each tangram piece be a different color so that the class can identify the location and changed orientation of each shape.

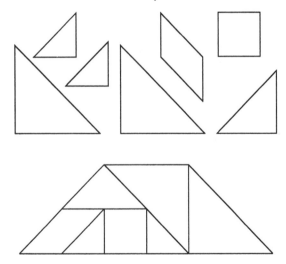

Tangram problem.

Sunflower Seed Lab Activity

An interesting laboratory activity, which gets at such important concepts as variation, frequency, range, and norm, involves the use of sunflower seeds. Give each pair of students about 200 seeds and 14 paper cups labeled 0 through 13. Instruct them to count the number of black stripes on both sides of each seed and then place each seed in the correspondingly numbered cup. Total class results are compiled by emptying the paper cups into 14 numbered hydrometer jars (borrow these from the science department) or tall, thin, olive jars lined up on the front desk. This activity is a golden opportunity to teach students the difference between the graphing of discrete and continuous data. The sunflower seed bar graph is a concrete example on which to base developmental questions and subsequent exercises on graph paper.

Selection 23

Teacher's Concept Map

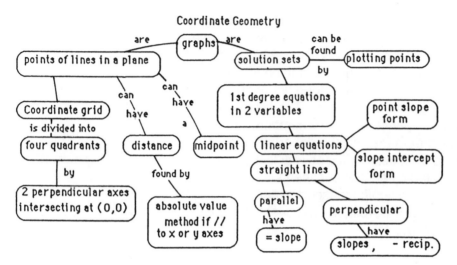

Part of a concept map on coordinate geometry. *Permission granted by David Laioso.

Implementing the NCTM Standards
© 1994, Janson Publications, Inc.

Selection 24

Learning Hierarchy

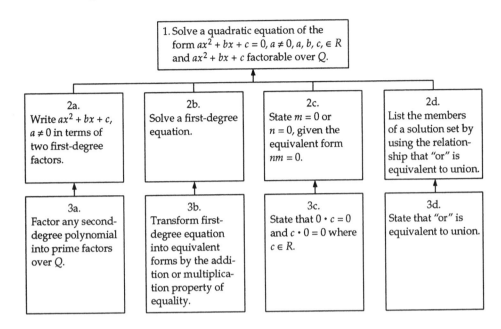

1. Solve a quadratic equation of the form $ax^2 + bx + c = 0$, $a \neq 0$, $a, b, c, \in R$ and $ax^2 + bx + c$ factorable over Q.

2a.
Write $ax^2 + bx + c$, $a \neq 0$ in terms of two first-degree factors.

2b.
Solve a first-degree equation.

2c.
State $m = 0$ or $n = 0$, given the equivalent form $nm = 0$.

2d.
List the members of a solution set by using the relationship that "or" is equivalent to union.

3a.
Factor any second-degree polynomial into prime factors over Q.

3b.
Transform first-degree equation into equivalent forms by the addition or multiplication property of equality.

3c.
State that $0 \cdot c = 0$ and $c \cdot 0 = 0$ where $c \in R$.

3d.
State that "or" is equivalent to union.

Selection 25

Course Planning Schema

Is a course plan, then, merely a list of unit plans designed to fit within a 140 day school year? There are basic flaws in that approach even when each individual unit has been thoughtfully designed. Yes, sequencing may be out of whack, interconnections may be ignored, and mindless redundancy may take the place of meaningful spiraling. There's only one way to avoid these disasters and that way begins with a thorough study of all available course materials. We recommend that you commit to paper the five to ten *major* instructional objectives of the course. If your list exceeds ten statements, you may be including some less important objective or ignoring the dependency of one objective on another. When you're satisfied that your list emphasizes the core of the course (Did you include processes as well as products of mathematics?), sketch in a simple learning hierarchy among just those five to ten major objectives. Don't worry if interconnections are rare. You may have identified several distinct and relatively independent terminal objectives. These statements, then, would become the uppermost objectives in a vast learning hierarchy. Next, match the topics, into which you had previously subdivided the course, with the objectives. If interconnections are missing here, go back to the drawing board. Finally, estimate the number of days to be assigned to each topic on the basis of (1) the relative contribution of each unit to the objectives, (2) the location of each unit in the sequence, and (3) the presumed range of student intellectual development. One teacher, Ms. Aronowitz, developed this kind of course outline for tenth-grade geometry. She wrote five major course objectives.

1. Construct a synthetic argument of a proof, given hypotheses from the major topic areas of plane geometry.

2. State generalizations in "If, then" form, given experiences reflecting a pattern.

3. Evaluate an argument (geometric or nongeometric) as to its match to deductive logic, its precision, and rigor.

4. Calculate measures (linear, angular, circular, area, volume) given data and relationships, diagrammatic or written, from which to obtain further data and/or rules.

5. Analyze the relationships between real-world phenomena and corresponding mathematical models from geometry.

Implementing the NCTM Standards
© 1994, Janson Publications, Inc.

Then she matched the five objectives to the eight unit areas into which she had previously divided the course (see the figure below). Notice the subdivisions under "Lines, planes, angles" and "Congruence." Constructions are listed under both unit topics, and coordinate geometry under unit topic 1. Subdivisions still must be added to the remaining units. The first two units were further synthesized by the addition of topic sentences. The beginning of proof considerations will be the focus of unit 1—for example, distinctions between assumption and theorem and between a geometric proof and a rationalization based on many experiences. However, the writing of synthetic arguments will be left for unit 2. That's also depicted by the growing width *and* length of the arrow associated with objective 1. Why does that arrow stop short of topic 8? Ms. Aronowitz is following well-established practice in eliminating this topic from those in which students will have to be able to write proofs. In like manner, each of the other arrows representing objectives identified the units in which that objective will be given explicit attention. See if you can follow Ms. Aronowitz's thinking here. The curved arrows connecting unit topics are other indicators of sequence. However, these represent the axiomatic links inherent in the connected topics rather than strands or overarching themes. "Parallelism" and "Area of polygons" both contain essential prerequisites to the proofs of the basic similarity theorems in the sequence being followed by Ms. Aronowitz so two curved arrows lead from unit topics 4 and 5 to topic 6. Each of the other curved links has a like reason for existing and, of course, multiple other links

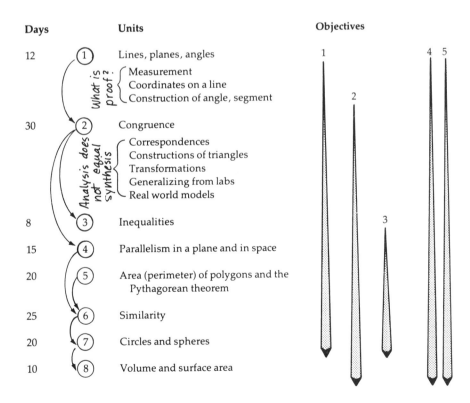

A course planning schema for tenth grade geometry.

are possible. But this teacher envisioned these as the major hierarchical connections which she should emphasize. A thorough grasp of the subject matter of geometry is clearly essential.

Now turn to the time allotments for each unit. Why were some planned for as little as 8 days while others were given as much as 30 days? Ms. Aronowitz identified "congruence" as the initial key to proof—the time to involve students in many concrete experiences, to build on the concepts and principles introduced in unit 1, and to develop the writing of simple synthetic arguments. On the other hand, "inequalities" would be built on prerequisites from other courses. The novel aspect would be the introduction of the indirect method of proof that would then become a strand throughout successive units. Other time allotments were arrived at by this kind of balancing of sequence, prerequisites, extent of novelty, and nature of objectives. There is nothing sacrosanct about these figures. They represent the best thinking of this teacher. As always, feedback during instruction may result in some differences between the actual and estimated times. Such information should be recorded and appropriate modifications made for future years. But any change must continue to reflect the conceptual framework designed by the teacher.

Now reflect on your own reactions to this particular overall course-planning guide. Does it provide a frame of reference for planning the sequence of the units it subsumes? Does it represent an accurate view of contemporary geometry and yet have high potential relevance for tenth-grade students? Are interrelationships among ideas within and across units classified? We answered yes to all these questions after studying the fully completed schema. Further, we judged it to have excellent potential for its intended purpose. Does this mean that this specific schema is *the correct one* to use for a tenth-grade geometry course? No, we would make no such claim—only that it ought to be a highly usable one for the person who created the schema and could implement it effectively.

Implementing the NCTM Standards
© 1994, Janson Publications, Inc.

Selection 26

Advance Organizers

It should be obvious from the word *advance* that an advance organizer comes prior to something else and from the word *organizer* that it is designed to facilitate putting things together in a meaningful way. Thus the term itself tells when an advance organizer occurs in an instructional sequence and sheds light on its overall purpose. David Ausubel (1963, 1968), the cognitive psychologist who originated this idea, has identified the attributes essential to this concept. First, *the advance organizer must present relevant content ideas that are of a higher order of abstraction, generality, and inclusiveness than the new material to follow.* Advance organizers typically take the form of broad concepts, rules/principles, thought models, theories, or conceptual schemes (themes) which subsume the more detailed knowledge to be learned next. Thus it becomes clear that teachers must have command of the structure of the subject matter, as presented in Chapter 4, if advance organizers are to be used in instruction. For example, the thought model of "group" could be taught as an advance organizer for related subsumed concepts/rules such as set, binary operation, closure, Abelian or commutative group, associative property, and permutation group.

Second, *the advance organizer must be presented in terms of what is already known by the learner.* In other words, the teacher must find out what relevant knowledge the learners possess and then use it to teach the generalization(s) that will serve as the advance organizer. For example, assume that the teacher finds out that the students already know (either from life experience or from previous instruction) that the order of operations makes a difference in some computations and activities (as in subtraction of natural numbers or in putting on one's socks and shoes). The teacher can then use these ideas as prerequisites to build comprehension of the desired generalization (group), which in turn will subsume the related facts, operational definitions, concepts, and rules/principles to be learned in the new work.

You should be starting to get the idea of what an advance organizer is. Now let's consider a few potentially confusing non-examples—that is, what an advance organizer *is not.* Many mathematics texts begin each chapter with a short introductory or overview section that consists of anywhere from one paragraph to several pages of writing. Often the one- or two-paragraph variety is really a summary of the main ideas to be treated in the chapter. These overviews almost always fail to meet *either* of the essential criteria for an advance organizer since (1) the ideas presented in capsule form are at the same level of abstraction as the content to follow and (2) little or no provision has been made to use the existing knowledge of the reader to teach the ideas summarized in such an introductory section. Similarly, several pages of introductory material tracing the historical development of a major mathematical idea may stimulate the interest of some students but fail to qualify as an advance organizer on the basis of one or both essential criteria. On the other hand, the history of mathematics, as well as the conceptual themes of mathematics, can be used to structure advance organizers....

Why are advance organizers important? One of their major functions is to facilitate the initial learning of new material so that it is of the meaningful variety (as opposed to rote). There is a growing body of research evidence that the degree of meaningfulness of newly learned material correlates positively with both remembering and the ability to use that material in applicable situations. Since both remembering (retention) and future applied use (transfer) are major goals of education, any device with potential for promoting their achievement merits our careful attention and best efforts.

Implementing the NCTM Standards
© 1994, Janson Publications, Inc.

Selection 27

Homework Post-Mortem

If a teacher wishes to ascertain the extent of student mastery of homework tasks, both of the following illustrations have proved useful.

1. In a class of 24, T sends 6 Ss to the board (at or prior to the opening bell) to put answers to 6 questions on the board *without* taking homework papers with them. T instructs Ss at seats to check their solutions against work as it is being put on board. T then tells the seated Ss that they will be required to talk through one of the examples, agree and tell why, correct errors and give reasons, or explain alternative correct solutions as soon as Ss at the board finish. Meanwhile, T walks around, makes a quick check of homework papers, identifies (a) those papers that are complete, (b) those on which various problems have been left out, and (c) types of errors made. Then T sends 6 *other* Ss to the board to critique the earlier work. Next, a third set of 6 students may be used to settle disagreements amoung the previous 2 Ss who worked at a problem.

or

2. T distributes copies with answers, or solutions and answers (overhead may be substituted for individual copies) *as Ss* enter the room. Ss are directed to check their work against the standard displayed by the copy or on the overhead while T walks around to check as in example 1 above. then T selects Ss to explain typical errors T has detected from papers.

If the teacher is satisfied that most students have achieved mastery of the task but wishes to provide feedback in an efficient way on details of the procedure, the teacher may:

1. Check papers as Ss walk in, identify correct answers, and instruct Ss with correct models to put work on board. Then, direct Ss to stay there to explain and to answer questions from other Ss.

or

2. Give selected Ss acetate sheets to take home. Then tell them to do the assigned problem on acetate for projection the next day. Also, inform them that they will be expected to explain and answer the questions of other Ss.

Selection 28

Student Concept Map

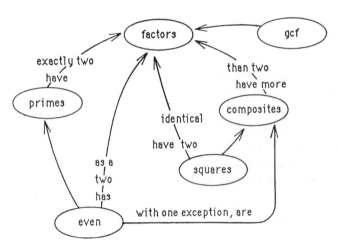

Concept map.

Notice the problem this student had with the concept EVEN. [The figure above] is one example of a concept map, a diagram that students construct to try to picture how they mentally link concepts.

Implementing the NCTM Standards
© 1994, Janson Publications, Inc.

Selection 29

Supervised Practice Sample

T has just completed a lecture and Q/A sequence on procedures for rewriting equations of functions in standard form. A straw vote on a single practice example showed most Ss with the correct answer. Therefore, T directs the Ss to begin working on the rest of the examples. Before T moves from the front, T checks to make sure that all Ss have started and tells a S whose hand is raised for help to reread notes until T has had a chance to check on others' progress. All others are now at work. T tours the room (as suggested in Chapter 1) and sees that only a few Ss are having difficulty with examples 5 and 8. T praises individual Ss quietly, indicates in a clear voice that many Ss are making good progress, quietly gives a cue to one S in difficulty, then moves to a second S and notices that some Ss have almost finished all nine examples. T sends selected Ss to the board, asks one S to serve as tutor for any S needing help, assists another S, and checks progress of class. (T has to decide if enough Ss are far enough along to stop the practice and look at the models of correct performance on the board. T must also decide if progress was sporadic or substantial. Do the Ss need additional practice on the same kind of examples or on a more complex set, or should T begin displaying a sequence of computer graphs so that Ss can check the equations in standard form against these graphs?) T calls the class to attention, tells them to compare their work with the board work and to be prepared to identify areas of difficulty. T stands in the rear quietly while this occurs and then takes straw votes on the number of Ss who agree or disagree with board work and seeks corrections where needed. Ss are frequently asked to respond to Q from other Ss.

Selection 30

Table of Specifications

What are you assessing? It should be the extent to which the students can meet the objectives of the unit. If your instruction was directed toward the mastery of objectives at levels I and II of the cognitive domain, then your *test results* should *report whether students did or didn't achieve each objective (criterion-referenced testing)*. If you included low-level objectives that all or most students should master as well as higher-level objective that fewer students might attain, you would be more likely to report *test results in terms of each student's relative position in the group (norm-referenced testing)*. Each of these types of testing is characterized by specific test construction principles. Since the secondary school teacher's test more often is modeled after norm-referenced instruments, we have chosen to examine testing from that vantage point. In either case, test planning begins with the consideration of the selected objectives.

Mr. Zidonis, an eleventh-grade teacher, began his test planning by completing the margins of what is called a *table of specifications*. He listed major content topics of the unit in the horizontal margin and the taxonomic levels of the cognitive objective in the vertical margin (see Figure 1).

	Objectives by Cognitive Level			
Units of Content	**I**	**II**	**III**	**Above III**
Basic probability concepts				
Multistage events				
Independent and dependent events				
Permutations and combinations				
Bernoulli Experiments				
Binomial Theorem				

Figure 1: A partially completed table of specifications.

Why not list each cognitive objective? Just imagine the size of the resulting table! Such a table of specifications would become an unmanageable monster rather than a guide to test design. Mr. Zidonis apparently included all levels of the cognitive taxonomy in his instruction, and he intends to sample the effectiveness of that instruction at each level. But levels of the affective domain and objectives in the psychomotor domain are not included. Did he ignore these domains in his instruction? Definitely not, but he chose to assess them by means of long-term projects and anecdotal

Implementing the NCTM Standards
© 1994, Janson Publications, Inc.

records during lab work rather than as part of the unit test. (Other ways of assessing objective in these domains are treated in later sections of this chapter.) Decisions! Decisions! There's no getting around it. The decisions that result in the completed margins of the table of specifications have already framed in the future test. But the professional testmaker wouldn't stop there. Next, each box of the table would be filled in with a decimal representing the instructional emphasis given the topic and the matching objectives. "You must be kidding!" Well, *they* aren't, but then their purpose and skills differ from those of the beginning teacher. Your next step is considerably different. You begin selecting and/or generating items.

Units of Content	Objectives by Cognitive Level				
	I	II	III	Above III	Credit
Basic probability concepts	3	1,6,17[3]		10[3]	12
Multistage events	12	4,11[3]	13[3]		10
Independent and dependent events	7	2,14[3]	15[3]		10
Permutations and combinations		9,16[3], 19[3],20a[5]	20b[5]		18
Bernoulli Experiments		5,8	21[10]		14
Binomial Theorem				18[3],22[10]	13
Credit Totals	6	34	21	16	77

Figure 2: A completed table of specifications.

References

Farrell, M.A. 1970. Area froma triangular point of view. *Mathematics Teacher* 63 (1): 18–21.

Farrell, M.A., and Farmer, W.A. 1985. Adolescents' performance on a sequence of proportional reasoning tasks. *Journal of Research in Science Teaching* 22 (6): 503–518

Karplus, R.; Lawson, A.E.; Wollman, W.; Appel, M.; Bernoff, R.; Howe, A.; Rusch, J.J.; and Sullivan, F. 1977. *Science teaching and the development of reasoning* (Volumes for Biology, Chemistry, Earth Science, General Science, and Physics). Berkeley, CA: The University of California.